Surface Engineering
Volume I: Fundamentals of Coatings

Surface Engineering
Volume I: Fundamentals of Coatings

Edited by

P.K. Datta and J.S. Gray

University of Northumbria at Newcastle, Newcastle upon Tyne

ROYAL
SOCIETY OF
CHEMISTRY

Based on the Proceedings of the Third International Conference on
Advances in Coatings and Surface Engineering for Corrosion and
Wear Resistance, and the First European Workshop on Surface Engineering
Technologies and Applications for SMEs. Both events were held at the
University of Northumbria at Newcastle, Newcastle upon Tyne, UK,
on 11–15 May 1992.

The front cover illustration was kindly provided by
Tecvac Ltd., UK.

Special Publication No. 126

ISBN 0-85186-665-4

A catalogue record for this book is available from the British Library

Published by The Royal Society of Chemistry,
Thomas Graham House, Science Park, Cambridge
CB4 4WF

Printed in Great Britain by Redwood Books Ltd., Trowbridge, Wiltshire

Preface

Surface Engineering is an enabling technology widely applicable to a range of industrial sector activities. It encompasses techniques and processes capable of creating and/or modifying surfaces to provide enhanced performance such as wear, corrosion and fatigue resistance and bio-compatibility. Surface engineering processes can now be used to produce multilayer and multicomponent surfaces, graded surfaces with novel properties and surfaces with highly non-equilibrium structures. In a broad sense surface engineering covers three interrelated activities:-

1. _Optimization of surface/substrate properties and performance_ in terms of corrosion, adhesion, wear and other physical and mechanical properties.

2. _Coatings technology_ including the more traditional techniques of painting, electroplating, weld surfacing, plasma and hypervelocity spraying, various thermal and thermochemical treatments such as nitriding and carburizing, as well as the newer processes of laser surfacing, physical and chemical vapour deposition, ion implantation and ion mixing.

3. _Characterization and evaluation of surfaces and interfaces_ in terms of composition and morphology, and mechanical, electrical and optical properties.

It is now widely recognized that the successful exploitation of these processes and coatings may enable the use of simpler, cheaper and more easily available substrate or base materials, with substantial reduction in costs, minimization of demands for strategic materials and improvement in fabricability and performance. In

demanding situations where the technology becomes
constrained by surface-related requirements, the use of
specially developed coating systems may represent the only
real possibility for exploitation.

The three volumes of *"Surface Engineering"* have been
prepared focusing attention on these comments and have
been principally based on papers presented at the 3rd
International Conference on Advances in Coatings and
Surface Engineering for Corrosion and Wear Resistance.
The structure and contents of the volumes in this series
have been conceived to provide a number of interrelated
themes and a coherent philosophy. Additional material has
been incorporated to complement the information delivered
at the Conference. As such, the text provides a useful
blend of Keynote, review, scientific and state-of-the-art
type papers by international authorities and experts in
surface engineering.

"Surface Engineering" is structured in three volumes.
The first volume, *Fundamentals of Coatings* considers
principles of coating/substrate design in high temperature
and aqueous corrosion and wear environments, and scans the
coatings' spectrum from organic, through metallic to
inorganic. Here there is a general emphasis on the
science and design of coating/substrate systems rather
than technology. The second volume, *Engineering
Applications*, is dedicated to topics concerning the
performance of coatings and surface treatments embracing
four main areas - the inhibition of wear and fatigue,
corrosion control, application of coatings in heat engines
and machining, and quality and properties of coatings.
Finally, the third volume has two main thrusts: *Process
Technology and Surface Analysis*. Both areas are clearly
central to surface engineering, and each holds particular
promise, not only for improvements in existing types of
coatings' performance, but also in the design, development
and evaluation of totally new - for example hybrid -
coating/substrate systems.

The editors wish to pay tribute to Dr. Tom Rhys-Jones who
recently died.

Contents

Section 1.5 Metallic Coatings

Section 1.6 Ceramic and Glass Ceramic Coatings

Contents
Volume II: Engineering Applications

Contents
Volume III: Process Technology and Surface Analysis

Acknowledgements

The editors wish to express their gratitude for the support extended by:

The Commission of the European Communities, The Department of Trade and Industry, The Institute of Materials, The Institute of Corrosion, Northern Electric, METCO Ltd., Cobalt Development Institute, LECO Instruments and Severn Furnaces.

Special thanks are due to Dr N E W Hartley of the CEC for his continual work and support of surface engineering.

The support and encouragement of Dr C Armstrong, Head of the Department of Mechanical Engineering and Manufacturing Systems, is gratefully acknowledged. Thanks are also due to Prof J Rear for opening the Conference.

The human commitment to any conference or book is substantial and often not fully acknowledged. In this final regard the work of Kath Hynes, Pauline Bailey, the technicians from the Department of Mechanical Engineering and Manufacturing Systems and the members of the Surface Engineering Research Group should be fully recognized.

Finally special commendation is reserved for Lee Comstock who administered the 3rd International Conference on Advances in Coating and Surface Engineering for Corrosion and Wear Resistance and the 1st European Workshop on the Application of Surface Engineering for SMEs, and was also responsible for the retyping of many conference papers.

P.K. Datta
J.S. Gray
University of Northumbria at Newcastle

Introduction

SURFACE ENGINEERING – THE NEW CHALLENGE IN MATERIALS
TECHNOLOGIES

I readily accept this opportunity to offer some person-
alized opinions on Surface Engineering.

Let me make it clear – it is not my aim to inform you
of the latest technical progress – indeed, the papers that
follow will attest to the developments in this new buoyant
theme in the materials sciences and technologies.

My principal aim is to affirm the critical importance
of this theme for modern industrial technologies. In
brief, I shall attempt to set this theme in its historical
and methodological context in the materials technologies
and with a glance at the wider horizon of potential growth
and impact on industry.

I do not know when the expression "surface
engineering" was first used to cover technologies relating
to surface modification; however, I do remember my old
friend Professor Tom Bell speaking volubly and excitedly
on this some 20 years ago. Many of you will remember that
surface treatment technologies such as shot-peening,
galvanizing, carburizing and nitriding were commonly
practised for many decades. In recent years, we have seen
a number of surface treatment techniques of a
sophisticated nature emerging from their position as
largely experimental laboratory techniques to achieve a
new status of commercial maturity. We will not go into

the details of the frustrating slowness of some of these developments, which reflect not only the difficulties in scaling up laboratory techniques, but also the general conservatism in the engineering industries and the reluctance to change from established habits. However, the expression "surface engineering" has now a wide acceptance and the word "engineering" fittingly describes the intention to induce changes to procure a designed end.

In this field, we are in the business of manipulating the properties of materials through the deposition of films and coatings or by changing the very physical and chemical character of the top atomic layers at the surface. The aim here is to procure a benefit through either improved performance in the conditions of operation or through extending the life of the component. Hence, surface engineering for structural materials can lead indirectly to the innovation of effectively new materials based on conventional materials. In the case of electronics or information technologies, we are presented with other vistas - fascinating possibilities where new materials with totally new physical and electronic properties can be synthesized using the techniques of surface engineering, such as epitaxial growth of very thin modulated layers on a surface.

My opinion is that surface engineering is rapidly becoming a highly significant and critical vector in the materials sciences and technologies, with considerable future industrial potential.

It is interesting to look back briefly into the short history of the development of the subject matters which now define materials science and technology and to glance at some of what we may describe as the landmarks in the subject matter.

The era of modern scientific physical metallurgy, the very bedrock of our materials science and technology today, is barely two-hundred years old. We cannot go into detail of how this subject emerged from a mix of disciplines such as chemistry, practical metallurgy, (and indeed alchemy) - however there is a landmark which symbolizes a break from the traditional, practical procedures used to understand the nature of steel and other alloys; this was when Professor Sorby in Sheffield turned his microscope from examining rock and mineral samples to the inspection of the microstructure in polished steel samples. Events issuing from this led to the emergence of modern microstructural metallurgy and the attempts to relate the physical and mechanical properties of metals to the crystalline structure.

In the intervening years, there have been a number of important landmarks: among these, in brief we quote:

- The emergence of the phase diagram and the application of thermo-chemical and thermo-dynamic principles for the interpretation of the equilibrium constitution of phases in alloy systems.

- The 1930s saw the development of the electron theory of metals and alloys and in particular, the rules for alloy formation, formulated by Professor Hume-Rothery.

- The rapid emergence of Organics as reliable engineering materials.

- In the post-war period there was considerable effort at relating mechanical properties to crystalline defects in structures.

- The development of Composites, both theory and practice, and its blossoming into its vigorous current state.

We now come to what I believe is a new landmark in the materials sciences and technologies, namely, Surface Engineering. It owes its origin to the recognition that for materials substances, the interface is the theatre of action - for chemical reactivity, for strength, for capillarity related processes. Here we note that, although surface engineering is conventionally used to refer to the interface between a bulk condensed solid and a vapour face, in fact the more generic term should be Interface Engineering, to include internal interfaces such as phase boundaries and grain boundaries, the important junctions where the body holds itself together and provides pathways for mass migration and shape changes. Remarkably, so much depends upon these almost two-dimensional zones of separation, which influence the very structural integrity of the bulk solid. Interfaces and interfacial engineering are today of enormous scope, touching many fields, including crystallization, joining processes and electronic device phenomena.

Free Surface Engineering is hence one aspect of the more general Interface Engineering. In itself, this is a vast field, its impact on properties is diverse and furthermore new techniques continue to provide tools for manipulating the surface properties. The feature which has brought surface engineering into great prominence today is the demonstrable improvement in properties and performance. One can extend the life of the component and this can be expressed in quantitative terms, helping to overcome the engineer's reluctance to change. Also the associated technologies are at the leading edge of industry in which there is scope for manoeuvre to introduce the techniques into commercial practice.

Among the methods of surface engineering, such as coating techniques, there is clearly a considerable industrial/economic potential for the protection of advanced materials during operations and for extending their working life, especially of valuable components. This is a view that is widely echoed by many competent R & D policy making bodies.

In general, coatings technologies are penetrating industrial manufacturing technologies on a very wide front. They offer technical solutions to diverse materials applications problems as well as innovative opportunities for new products. The applications range widely, for example: optical as well as aesthetic properties, chemical resistance such as corrosion protection, hardness improvements, wear and erosion resistance, procuring specific electrical and magnetic properties, improved high temperature performance as well as improved catalytic performance. The economic potential of coatings technologies has been widely recognized and is underlined in a study for the German BMFT - Table i illustrates well this optimistic forecast.

At this point, a word on the actual coating techniques. Clearly it is not our intention here to go into the multitude and diversity of specific techniques

<u>Table i</u> Industrial turnover and growth rates in coatings
technologies (Germany)

	10^9DM p.a.	% Growth p.a.
PVD	3.40	15
CVD	0.96	25
P-CVD*	0.05	40
Implantation	0.80	28

* Plasma-supported CVD

that are developing rapidly. We should recognize that
there does not exist a <u>best</u> coating process or facility –
for each problem, for each substrate/coating system, for
each application and each component type, a specific
technique and facility can be devised which is the most
appropriate. We note that the properties of the coatings
can depend significantly on the process mode.
Furthermore, there is a growing trend to use hybrid
techniques, that is, a combination of various processes to
produce the desired end product.

For our purposes it is sufficient to make the
following general classification in techniques for surface
modification; these are:

–   <u>Transfer Droplets.</u> Here and in particular with the
    mode of <u>low pressure plasma spraying</u>, there is the
    advantage of being able to deposit any material
    available in powder form and having a stable melting
    phase. This method has a high deposition rate and is
    most suitable for relatively thick coatings or free
    standing shapes.

–   <u>Transfer of Atoms or Ions.</u> This group of techniques
    comprises essentially CVD techniques (<u>chemical vapour</u>
    <u>deposition</u>) which can be applied as active or
    reactive processes, and PVD techniques (<u>physical</u>
    <u>vapour deposition</u>) such as deposition by evaporation
    or sputtering which can also be performed in a direct
    or reactive way by <u>ion implantation</u>. To improve the
    coating properties, there are efforts devoted to

combine CVD and PVD techniques with plasma or laser application.

- Application of Laser Beams. This refers to the burgeoning field of laser beam application which can be performed with or without the addition of foreign material.

We have already indicated that one of the reasons for the rapid development of surface engineering technologies is the demonstrable improvement in the life of the component for a particular end use. It has been said that only about 5% of the potential application areas for protective coatings are currently being exploited. Here, although cost is seen generally as the main restraint in market penetration, this is being challenged by the efforts of coating equipment manufacturers to improve and reduce the cost of their facilities as well as to seek for greater automation. With these cautionary remarks, there exist nevertheless many areas of application where considerable life improvement has been demonstrated. A selection of data is indicated in Table ii for the techniques of CVD, PVD and Ion Implantation.

Table ii Examples of life improvement in tools following surface treatments

Application	Method	Factor of Life Improvement
Drill	PVD	6-7
Abrasive nozzle	PVD	6
Drawing die	CVD	75
Mining collar	CVD	5
Injection moulding tool	Ion Implantation	3
Slitters for rubber	Ion Implantation	12
Thread cutting high speed steel	Ion Implantation	9

The data here are the output of much experimentation. Clearly, depending upon the article and its use, some remarkable improvements in life have been measured, some in the range of one to two orders of magnitude. This is as far as the materials technologists can go to demonstrate the potential of the technology. The critical decision as to whether to take on board the technology is of course up the industry itself; it requires the local evaluation on technico-commercial grounds of the cost/benefit factor in shifting into the new technology.

Above, I indicated that the potential for industrial exploitation of surface engineering is immense. Let me just mention briefly a recent list of materials requirements put forward by a panel of international petrochemical industries including the offshore industry, refineries, and general mining industry. Of the long list of requirements for new materials technologies, many of the improvements that have been demanded relate to the exploitation of surface engineering, viz.

- for centrifugal pumps: longer life impellers and wear rings

- for gas turbines: improved life of the hot gas part component

- for mining equipment and oil field pumps in sandy conditions: the development of erosion resistant coatings that "bend but do not break"

- for reciprocating compressors and plunger pumps: better abrasive wear resistant hard coatings

- for reciprocating engines: better low friction coatings which are wear resistant

- for hydraulically lubricated machinery: more reliable metallic coatings and platings for repairing rotor journal surfaces

- for pumps: more reliable bearings for sealless pumps and effort to eliminate the lubrication of journal bearings.

As for many other sciences and technologies, surface engineering as a discipline was identified first in Europe. We note, for example, that Harwell was probably the first to develop extensively ion implantation methods for improving corrosion and abrasion resistance in alloys. The recognition of surface or interface engineering as a

landmark theme is not only a European matter; it has also
been taken up actively in Japan and U.S.A. In particular,
micro-electronics and information technology have fuelled
aspects of surface engineering such as ion beam
engineering.

Europe is aware of the enormous importance and
potential of surface engineering. Various institutes in
Europe are responding to this challenge: thus for
example, the Institute for Advanced Materials of the Joint
Research Centre has installed a comprehensive Laser/Ion
Foundry in Ispra and an Advanced Coatings Centre in
Petten. However, the problem is a recurring one in Europe
- quick to develop the idea but slow to implement it; when
this is done, it is usually in a highly fragmented manner.
This has been the case for many areas of research.
However, efforts are being made to try to introduce some
measure of collaboration, in order to overcome duplication
of effort and put together supercritical activity fields.
This has been a corner stone of the European Commission's
R & D policy; that is, to introduce the instruments and
encouragement for the various teams scattered throughout
the Member States of the European Communities to try to
put together their effort into coordinated units of
supercritical size.

I believe that this holds for surface engineering,
and although this field has been supported in cost shared
activities under the Framework Programmes, there is a need
for more specific and targeted activities which could be
conducted by R & D networking procedures. Here, certain
laboratories might constitute nodal points in the Network
and could contribute according to their own intrinsic
capabilities in a complementary manner to the work of the
Network.

Before concluding, let me quote from a recent glossy
handout produced by an organization in Japan, known as
NEDO, which has recently set up an ion engineering centre
with a large capital outlay in which it is expected to
carry out research on new ideas in ion-assisted
technology. Here it is claimed that through ion
engineering one will combine metals, semiconductors,
ceramics, and organics to fabricate new materials. Since
it holds "infinite possibilities", research and
development in this technology "is being watched with keen

interest throughout the world". The unusual motto of this international centre is "to be healthy, be beautiful, be civilized and be original".

Although it is highly unlikely that we in Europe would approach this challenge in the same centralized manner, the message is there - interface engineering is coming of age and there is enormous excitement. Although Europe will not adopt the same motto as the Japanese, the European motto might be along the lines "interfacial engineering, Europe puts its act together".

E.D. Hondros
Institute for Advanced Materials
Commission of the European Communities
Petten, The Netherlands
(from the opening address of the 3rd International Conference on Advances in Coatings and Surface Engineering for Corrosion and Wear Resistance)

I am grateful to Dr. E. Lang of this Institute for his help in preparing material for this paper.

Part 1: Fundamentals of Surface Coatings

Section 1.1 High Temperature Corrosion

1.1.1
Estimating and Calculation of the Values of Thermodynamic Functions for Certain Group V and Group VI Metal Chlorides

K. N. Strafford,[1] G. Forster,[2] and P. K. Datta[3]

[1] DEPARTMENT OF METALLURGY, GARTRELL SCHOOL OF MINING, METALLURGY AND APPLIED GEOLOGY, UNIVERSITY OF SOUTH AUSTRALIA, THE LEVELS, SOUTH AUSTRALIA, 5095

[2] METALTECH SERVICES LIMITED, COUNTY DURHAM DH8 9HU, UK

[3] SURFACE ENGINEERING RESEARCH GROUP, UNIVERSITY OF NORTHUMBRIA AT NEWCASTLE, NEWCASTLE UPON TYNE NEI 8ST, UK

1   INTRODUCTION

Halides, in particular the chlorides, are used extensively in many metallurgical process routes[1]. The use of chloridizing roast to convert pyrite into a water soluble product using a reaction of the type

$$MeS + 2NaCl + 2O_2 \rightleftharpoons Na_2SO_4 + MeCl_2 \qquad (1)$$

is one such example. High temperature chloridizing roasts, where non-ferrous metals are converted into volatile chlorides by the use of elemental chlorine are well-established, typical examples being:

$$MeO + Cl_2 \rightleftharpoons MeCl_{2(g)} + 1/2O_2 \qquad (2)$$

$$MeS + Cl_2 + O_2 \rightleftharpoons MeCl_{2(g)} + SO_2 \qquad (3)$$

where Me is a divalent metal.

Some of the reasons for the use of halides in the production of reactive metals are:

(i) the relatively high vapour pressures of halides, allowing the possibility of purification by distillation;

(ii) the low solid solubility of chlorides in metals, permitting the production of pure metals.

In addition, for certain metals:

(iii) the low melting points of chlorides, coupled with high electrical conductivities, allowing the possibility of electrolytic reductions;

(iv) the utilization of the aqueous solubility property facilitating purification by crystallization.

Table 1   A selection of high temperature process
           environments involving chlorination[2]

HIGH TEMPERATURE PROCESS	TEMPERATURE RANGE	ENVIRONMENT
Mineral chlorination: - precious metals; Ti and Zr production. Chlorinators.	300-900 °C	Chlorine
High temperature chlorination: reactors.	Up to 550 °C	Chlorine
Titanium oxide production (chlorine route) Heater tubes in $TiCl_4$ circuit.	900 °C +	Oxygen and chlorine
Aluminium melting. Mg removal by Cl_2 injection	Up to 850 °C	Flue gases with Cl_2, HCl, S etc.
Fibreglass manufacture: include recuperators.	Up to 900 °C	Flue gases, chlorine and sulphur species
Waste incineration: incineration of plastic waste-refuse as supplement to existing fossil fuel.	Up to 900 °C	Variable flue gases, including chlorine and sulphur species
Uranium refining: hydrofluorinator and fluorinator reactors.	600 °C	HF Fluorine
Reprocessing of fuel elements: hydrofluorinator.	650 °C	Molten fluoride HF
Plant for H_2 from water (thermochemical reactions): nuclear/solar heat.	Up to 900 °C	HCl
Power generation - boilers - coal fired; boilers - oil-fired: superheater supports, gas turbines.	Up to 950 °C	Combustion gases including HCl, fuel ash sulphate/chloride
Fluidized bed combustor: heat exchanger tubes.	850 °C	HCl from coal with $CaSO_4$, CaO (sulphur/oxygen/chlorine)

<u>Table 2</u>   Free energy data for Group V and VI metal
chlorides (constants A,B and C in Expression:
G = A + BT - CTlnT)

REACTION	ΔG^o in calories			Temp. Range (K)
	A	B	C	
$<Mo> + (Cl_2) = <MoCl_2>$	-70407	-4.01	58.48	298 - 803
$2/3<Mo> + (Cl_2) = 2/3<MoCl_3>$	-62903	-3.8	61.64	298 - 1290
$2/3<Mo> + (Cl_2) = 2/3\{MoCl_3\}$	-79023	-10.43	114	1290 - 1500
$1/2<Mo> + (Cl_2) = 1/2<MoCl_4>$	-58491	-4.8	68.80	298 - 590
$1/2<Mo> + (Cl_2) = 1/2\{MoCl_4\}$	-60114	-12.75	115.89	590 - 680
$1/2<Mo> + (Cl_2) = 1/2(MoCl_4)$	-63969.6	-13.52	126.43	680 - 500
$2/5<Mo> + (Cl_2) = 2/5<MoCl_5>$	-37146	-3.75	51.39	298 - 467
$2/5<Mo> + (Cl_2) = 2/5\{MoCl_5\}$	-42055	-9.98	106.23	467 - 537
$2/5<Mo> + (Cl_2) = 2/5(MoCl_5)$	-48055	-9.21	112.35	537 - 1500
$1/3<Mo> + (Cl_2) = 1/3<MoCl_6>$	-41667	-3.28	50.52	298 - 527
$1/3<Mo> + (Cl_2) = 1/3(MoCl_6)$	-42603	-2.61	52.27	527 - 1500
$1/2<Nb> + (Cl_2) = 1/2<NbCl_4>$	-81552	-3.80	61.11	298 - 728
$1/2<Nb> + (Cl_2) = 1/2(NbCl_4)$	-85046	-3.10	61.29	728 - 1500
$2<NbCl_4> = <NbCl_3> + <NbCl_5>$	28300	-	-47.0	475 - 590
$2/5<Nb> + (Cl_2) = 2/5<NbCl_5>$	-76200	-3.71	59.61	298 - 483
$2/5<Nb> + (Cl_2) = 2/5\{NbCl_5\}$	-82445	-10.42	113.72	483 - 523
$2/5<Nb> + (Cl_2) = 2/5(NbCl_5)$	-87685	-9.65	118.01	523 -
$2/5<Ta> + (Cl_2) = 2/5<TaCl_5>$	-83535	-3.74	59.94	298 - 590
$2/5<Ta> + (Cl_2) = 2/5\{TaCl_5\}$	-88169	-10.48	112.68	590 - 611
$2/5<Ta> + (Cl_2) = 2/5(TaCl_5)$	-92577	-9.74	115.27	611 -
$(TaCl_5) = (TaCl_3) + (Cl_2)$	33600	-	-31.3	-
$2<TaCl_4> = (TaCl_3) + (TaCl_5)$	27000	-	-48.3	-
$3<TaCl_3> = 2(TaCl_4) + (TaCl_5)$	33400	-	-32.3	-
$2/3<V> + (Cl_2) = 2/3<VCl_3>$	-89333	-2.84	57.0	298 - 698
$2<VCl_3> = <VCl_2> + <VCl_4>$	38000	-	39.6	698 - 928
$1/2<V> + (Cl_2) = 1/2<VCl_4>$	-68966	-	72.3	0 - 248
$1/2<V> + (Cl_2) = 1/2\{VCl_4\}$	-67608	-5.00	72.3	248 - 425
$1/2<V> + (Cl_2) = 1/2(VCl_4)$	-56508	-8.00	72.3	425 - 2190
$<V> + (Cl_2) = <VCl_2>$	-98000	-3.03	50.2	298 - 1250

Whilst the properties of high vapour pressures and low melting points of chlorides can be utilized beneficially in extraction metallurgy, the same physical-chemical properties can also promote degradation in structural materials when exposed to chlorine-containing environments at elevated temperatures. Many established industrial processes contain gaseous halogen environments where operating temperatures commonly exceed 500°C and where plant conditions often involve environments containing oxidant species such as steam, air or sulphur in conjunction with one or more of the halogens. Examples are listed in Table 1, after Elliot[2].

Of primary importance in all these situations is to know, or the ability to calculate, the thermodynamic parameter which provides a measure of the extent to which a reaction between specified substances can occur under a given set of conditions - the Gibbs' free energy function G - defined by the relationship:

$$\Delta G = \Delta H - T\Delta S \qquad (4)$$

For reactions in which the standard-state condition is satisfied, the symbols become, for a specified temperature T,

$$\Delta G^{o}_{T} = \Delta H^{o}_{T} - T\Delta S^{o}_{T} \qquad (5)$$

where ΔG^{o}_{T} is the standard free energy change.

Calculations of ΔG^{o}_{T} for reactions of the type $2xMe + yCl_2 = 2Me_xCl_y$, over a wide range of temperature are extremely important, since it is often necessary to take a rigorous account of the temperature variation of ΔG^{O}.

2   DERIVATION AND CALCULATION OF THERMODYNAMIC DATA

The free energies of formation of selected inorganic chlorides were first compiled by Villa[3]. The variation in Gibbs' free energy with temperature for the chlorides of some metals considered important to alloy design for structural engineering components has been determined. Such data for chromium, cobalt, iron, nickel, aluminium and titanium have been calculated from established thermodynamic data[4].

Figure 1 has been constructed using data obtained from Kubaschewski et al. [4,5] and Smithells[6]. Free energy data for the chlorides of molybdenum, niobium, tantalum and vanadium however are not readily available, but an attempt has been made to calculate ΔG^{o}_{T}, using well-established thermodynamic approximations. These results are shown in Figure 2, and the method of calculation, by way of example, is included in the Appendix to this paper: Table 2 lists the constants obtained from such calculations, then used to estimate the ΔG^{o}_{T} values quoted.

<u>Figure 1</u>   Variation of ΔG° with temperature for
chlorides of Al, Co, Cr, Fe, Ni and Ti

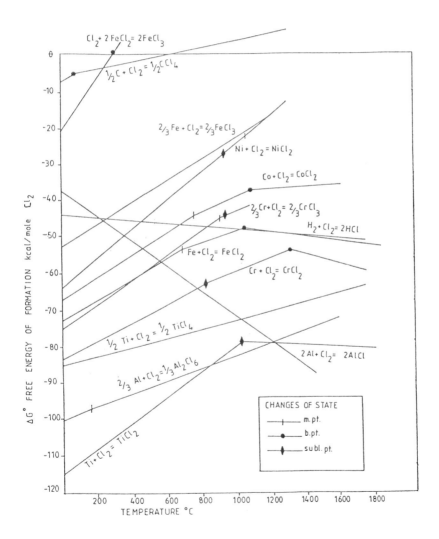

Figure 2   Variation of ΔG° with temperature for the
chlorides of Nb, Mo, Ta and V

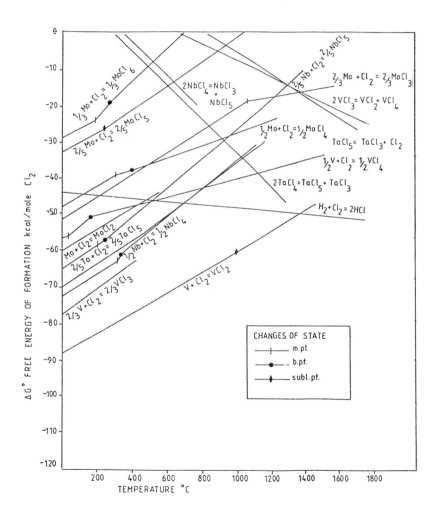

Unlike the situation with metal oxides, when carbon can be used as a reducing agent, it is to be noted that carbon tetrachloride is itself extremely unstable; however, hydrogen chloride becomes increasingly stable with increasing temperatures, so enabling hydrogen to reduce many metal chlorides at higher temperatures. The variations of $\Delta G°_T$ with temperature for hydrogen chloride are accommodated in both Figures 1 and 2. The accuracy of such calculated values will obviously depend upon the quality of the source data available. In all cases, the most accurate, or that which is assumed to be the most accurate, has been used. In the course of such calculations experimental errors will accumulate. Obviously, calculations employed should include the minimum number of steps to avoid error accumulation. Clearly, actual determination of heats of formation and C_p data for these metal chlorides of interest would reduce greatly the final error. Perhaps, cautious values of error, lying between $\pm$ 5 kcal could be applied, this giving at worst $\pm$ 25% for $1/3Mo + Cl_2 \rightleftarrows 1/3MoCl_6$, and, at best $\pm$ 7% for $V + Cl_2 \rightleftarrows VCl_2$.

Although the method of estimation used is thus open to conjecture, and criticism, it does provide the experimentalist and process engineer with a reasonable basis on which to commence research and project studies.

REFERENCES

1. T. Rosenquist, Principles of Extractive Metallurgy, McGraw-Hill, Inc, 1974.
2. P. Elliot, 'High Temperature Alloy Corrosion by Halogens', The Journal of Metals, July 1985.
3. H. Villa, 'Thermodynamic Data of the Metallic Chlorides', J. Soc. Chem. Ind. Supplementary Issue, No.1, 1950.
4. O. Kubaschewski, E. L. L. Evans, and C. B. Alcock, Metallurgical Thermochemistry, 4th Edition, Pergamon Press, London, 1967.
5. O. Kubaschewski and E. L. L. Evans, Metallurgical Thermochemistry, Butterworth-Springer, London, 1951, and Pergamon Press, London, 1958.
6. C. J. Smithells, Metals Reference Book, Vol. 2, Butterworths, London, 1955.

APPENDIX : CALCULATION OF ΔG^o_T USING THERMODYNAMIC APPROXIMATIONS

Values of ΔG^o_T can be calculated from standard values with certain
transformations using the procedure.

From $\quad \Delta G^o_T = \Delta H_T - T.\Delta S_T$
ΔG^o_T can be expressed as:

$$= \Delta H_{298} + \int_{298}^{T} \Delta C_p.dT - T.\Delta S_{298} - \int_{298}^{T} \frac{\Delta C_p}{T}.dT$$

$$= \Delta H_o + \int_{0}^{T} \Delta C_p.dT - T.\Delta S_0 + \int_{0}^{T} \frac{\Delta C_p}{T}.dT$$

$$= \Delta H_0 + \int_{0}^{T} \Delta C_p.dT - T\Delta S_0 - T\int_{0}^{T} \frac{\Delta C_p}{T}.dT$$

where

$$H_0 = H_{298} - \int_{0}^{298} \Delta C_p.dT$$

and

$$\Delta S_0 = \Delta S_{298} - \int_{0}^{298} \frac{\Delta C_p}{T}.dT$$

Example: The Formation of Niobium Pentachloride

Consider the reaction 2/5 Nb + Cl$_2$ $\rightleftharpoons$ 2/5 NbCl$_5$, and the derivation of
ΔG as f(T) line.
For this reaction strictly there is a need to consider,over a extended temperature range, three
variations, viz:

$$\text{2/5 Nb + Cl}_2 \rightleftharpoons \text{2/5 NbCl}_5\text{(s)}$$

and $\quad$ 2/5 Nb + Cl$_2$ $\rightleftharpoons$ 2/5 NbCl$_5$(l)

and $\quad$ 2/5 Nb + Cl$_2$ $\rightleftharpoons$ 2/5 NbCl$_5$(v)

each involving changes of state.

For any given reaction $\Delta Cp = \Sigma Cp$ (products) - ΣCp (reactants)

Cp data from sources are presented in the form:

$\quad$ Cp = a + 10^{-3} bT + 10^5 g T^{-2}
For Cl$_2$, Cp = 8.82 + 0.06 x 10^{-3} T - 0.68 x 10^5 T^{-2}; for
$\quad$ Nb , Cp = 5.885 + 0.96 x 10^{-3} T

However, for $NbCl_5$ no Cp data is available from the common sources of reference (4,6) but an estimate can be made on the basis of well-established empirical rules as follows, depending on the state of the $NbCl_5$ - solid, liquid or vapour.

1. For **solids,** Dulong and Petit's Law gives heat capacity per g. atom as around 6.2 cal. deg $^{-1}$ at room temperature, rising to about 7.25 cal deg $^{-1}$ at the first transition temperature.

2. The heat capacities of **liquids** are only slightly higher than those of solids near the melting point - by around 0.2 cal g.$^{-1}$ atom i.e. 7.45 cal. deg $^{-1}$.

3. For non-linear polyatomic **gases, or vapours** there is difficulty in predicting their heat capacities accurately. However an estimate may be obtained by adding 6 cal/deg (or 5 for a linear molecule) to four times the number of interatomic linkages in the molecule, Z, (irrespective of whether they are single or multiple bonds) e.g. for $NbCl_{5(g)}$

$Cp = (4 \times 5) + 6 = 26$ cal.deg $^{-1}$

If a value of "X" is assigned initially for Cp_{NbCl5}, then ΔCp for the reaction $2/5\ Nb + Cl_2 \rightleftharpoons 2/5\ NbCl_{5(s)}$

is : $2/5\ X - 11.17 - 0.44 \times 10^{-3}T + 0.68 \times 10^5 T^{-2}$

For convenience, letting $a = 11.17$

$$b = 0.44 \times 10^{-3}$$

and $g = 0.68 \times 10^5$

then Cp for the reaction $2/5\ Nb + Cl_2 = 2/5\ NbCl_5$

is : $2/5\ X - a - bT + gT^{-2}$

Inserting these appropriate values for X ($NbCl_5$ as solid), a, b, g and T gives an entropy value of 20.73 cal. deg^{-1}. mol^{-1}.

$$\Delta S_{298} = 21.6 - 53.3 - 3.5 = -35.2 \text{ cal.mol.}^{-1}\text{deg.}^{-1}$$

Whence $\Delta S_o = -35.2 - 20.7 = -55.9$ cal. mol.$^{-1}$ deg.$^{-1}$

Then: $\Delta G_a(T < 478) = -76200 + \int_0^T \Delta Cp\, dT - T\Delta S_0 - T\int_0^T \dfrac{\Delta Cp}{T}\, dT$

$$= -76200 + \left(14.88T - 11.17T - \dfrac{0.44}{2} \times 10^{-3}T^2 - 0.68 \times 10^5 T^{-1} \right)$$

$$- T(-55.9) - -14.88T \ln T + 11.17T \ln T + 0.44 \times 10^{-3}T^2$$

$$+ \dfrac{0.68 \times 10^5}{2} T^{-1}$$

i.e. $\Delta G_a = -76200 + 59.61T - 3.71T \ln T + 0.22 \times 10^{-3}T^2 - 0.34 \times 10^5 T^{-1}$

Calculations of ΔG_T^0 above the melting point of NbCl$_5$ must include the entropy and enthalpy of fusion.

For the transition 2/5 NbCl$_5$(l) $\rightleftharpoons$ 2/5 NbCl$_5$ (s) : let ΔG_b be the free energy change (for transition (b))

Since $\Delta H_{f(478)}$ = 6900 x 2/5 cal = 2760 cal

Then $\Delta H_{(478)}\dfrac{2}{5}NbCl_{5(l)} \rightarrow \dfrac{2}{5}NbCl_{5(s)} = -2760$cal

and $\quad \Delta S_{478} = -\dfrac{2760}{478} = -5.77 \quad$ cal. mol^{-1}deg.$^{-1}$.

Since $\quad \Delta Cp \quad$ for(b) $= \dfrac{2}{5}X - 11.71 - 0.44 \times 10^{-3}T + 0.68 \times 10^{5}T^{-2} \quad$ where $\quad X = 7.45 \times 6 = 44.7$

Then $\displaystyle\int_0^T \Delta C_p dT = 17.88T - 11.17T - \dfrac{0.44 \times 10^{-3}T^2}{2} - 0.68 \times 10^5 T^{-1}$

and $\displaystyle\int_0^T \dfrac{\Delta C_p.dT}{T} = \dfrac{2}{5}(44.7)\ln T - 11.17\ln T - 0.44 \times 10^{-3}T - \dfrac{0.68 \times 10^5}{2.T^2}$

Now
$$\Delta H_o = \Delta H_{478} - \int_0^{478} \Delta C_p.dT = -2760 - \int_0^{478} \Delta C_p dT$$

$-2760 - 6.71T - \dfrac{0.44 \times 10^{-3}T^2}{2} - 0.68 \times 10^5 T^{-1}$

$= -2760 - 3207.4 - 49.2 - 228.2$

$= -6244.8 \quad$ cal. mol^{-1}

and

$$\Delta S_o = \Delta S_{478} - \int_0^{478} \dfrac{\Delta C_p dT}{T}$$

$= -5.77 - 6.71T \ln T - 0.44 \times 10^{-3}T - \dfrac{0.68 \times 10^5}{2.T^2}$

$= -47.43 \quad$ cal. deg^{-1} mol^{-1}.

Now,
$$\Delta G_b = \Delta H_o + \int_0^T \Delta C_p dT - T\Delta S_o - T\int_o^T \dfrac{\Delta C_p dT}{T}$$

$= -6244.8 + 6.71T - \dfrac{0.44 \times 10^{-3}T^2}{2} - 0.68 \times 10^5 T^{-1}$

$-T(-47.4) - 6.71T \ln T + 0.44 \times 10^{-3}T^2 + \dfrac{0.68 \times 10^5 T^{-1}}{2}$

i.e. ΔG_b = -6244.8 + 54.11T - 6.71lnT + 0.22 x 10^{-3}T^2 - 0.34 x 10^{5}T^{-1}

Now for the reaction $2/5\ Nb_{(s)} + Cl_{2(g)} \rightleftharpoons 2/5\ NbCl_{5(l)}$, **(reaction c)**

ΔG_c is found by addition of equations ΔG_a and ΔG_b : i.e. $\Delta G_c = \Delta G_a + \Delta G_b$

For any temperature at T > 478 (but < 523 K)

$$\Delta G_c = (-76200 + 59.61T - 3.71T \ln T + 0.22 \times 10^{-3}T^2 - 0.34 \times 10^5 T^{-1})$$

$$+ (-6244.8 + 54.11T \ln T + 0.22 \times 10^{-3}T^2 - 0.34 \times 10^5 T^{-1})$$

i.e. ΔG_c = **-82445 + 113.72T - 10.42T ln T + 0.44 x 10⁻³ T² - 0.68 x 10⁵T⁻¹**

NbCl₅ boils at 250 ° C (523 K) and the free energy change associated with this transition
$2/5\ NbCl_{5(s)} \rightleftharpoons 2/5\ NbCl_{5(l)}$ **(Transition (d))**
has to be recognised.

Using Hess's Law ΔH_{298} for the reaction $2/5\ Nb + Cl_2 = 2/5 NbCl_{5(s)}$ is given by :

$$2/5 \Delta H_{NbCl_5} - \Delta H_{Cl_2} - 2/5 \Delta H_{Nb}$$
$$\Delta H_{298} = -7600 \quad \text{cal.mol}^{-1}.\text{deg}^{-1}.$$

To find ΔH_o by Kirchoff's Law,

$$\Delta H_{298} = \Delta H_o + \int_0^{298} \Delta Cp \,.dT$$

or, $\Delta H_0 = -76200 - \int_0^{298} \Delta Cp \,.dT$

But $\Delta Cp = 2/5X - \alpha - \beta T + \gamma T^{-2}$

thus, $\Delta H_0 = -76200 - \left[\dfrac{2}{5}XT - \alpha T - \dfrac{\beta T^2}{2} - \gamma T^{-1} \right]_0^{298}$

Here using the empirical ruling of Dulong and Petit, X attains a value of
6 x 6.2 = 37.2 cal. mol deg⁻¹.; inserting the values for X, α, β, γ and T yields :

$$\Delta H_0 = -76200 - \frac{2}{5} \times 37.2 \times 298 + (11.17 \times 298) + \frac{0.44}{2} \times 10^{-3} \times 298^2 + \left(0.68 \times \frac{10^5}{298} \right)$$

From which ΔH_0 = -74396 cal. mol ⁻¹.
The value of ΔS_0 is calculated thus

$$\Delta S_0 = \Delta S_{298} - \int_0^{T=298} \frac{\Delta Cp}{T} dT$$

Now, $\displaystyle\int_0^{T=298} \frac{\Delta Cp}{T} dT = \int_0^{298} \frac{2X}{5T} - \frac{\alpha}{T} - \frac{\beta T}{T} + \frac{\gamma}{T^3} dT$

$$= \left[\frac{2}{5}X \ln T - \alpha \ln T - \beta T - \frac{\gamma T^{-2}}{2} \right]_0^{298}$$

For the transition (d) $2/5 \, NbCl_{5(v)} = 2/5 \, NbCl_{5\,(l)}$, the heat of vapourisation

$$\Delta H_v(523) = \frac{2}{5} \times 13100cal = -5240cal.$$

and therefore $\Delta S_{523} = -5240/523 = -10.1$ cal.deg^{-1} mol^{-1}.

Since
$$\Delta Cp \quad for \quad (d) \quad = \frac{2}{5} \times 26 - 11.17 - 0.44 \times 10^{-3}T + 0.68 \times 10^5 T^{-2}$$

then
$$\int_o^T \Delta Cp \, dT = -0.77T - 0.22 \times 10^{-3}T^2 - \frac{0.68 \times 10^5 T^{-1}}{2}$$

and
$$\Delta H_0 = \Delta H_{523} - [-0.77T - 0.22 \times 10^{-3}T^2 - 0.68 \times 10^5 T^{-1} \;]_0^{523}$$

$$= -5240 - (-402.71 - 60.17 - 130) = -4647.1 \quad cal.$$

and
$$\int_0^T \frac{\Delta Cp}{T} dT = -0.77 \ln T - 0.22 \times 10^{-3}T - \frac{0.68 \times 10^{-5}}{2} \times T^{-2}$$

Hence
$$\Delta S_0 = \Delta S_{523} - \int_0^{523} \frac{\Delta Cp}{T} \, dT$$

$$= -10.1 - (-4.81 - 0.11 - 0.12)$$

$$\Delta S_0 = -5.06 \quad cal. \; mol^{-1}.deg.^{-1}$$

$$\Delta G_d = -5240 + 4.29T + 0.77 \ln T + 0.22 \times 10^{-3} T^2 - 0.34 \times 10^5 T^{-1}$$

Hence for any temperature $T > 523$

$$\Delta G_e = \Delta G_c + \Delta G_d$$

$$\Delta G_e = -5240 + 4.29T + 0.77T \ln T + 0.22 \times 10^{-3}T^2 - 0.34 \times 10^5 T^{-1}$$

$$-82445 + 113.7T - 10.42T \ln T + 0.44 \times 10^{-3}T^2 - 0.68 \times 10^5 T^{-1}$$

$$= -87685 + 118.01T - 9.65T \ln T + 0.66 \times 10^{-3}T^2 - 1.02 \times 10^5 T^{-1}$$

Plotting the equation ΔG_e against absolute tempertature yields a line which is essentially straight (see Figure 2) and because of the small contributions of the terms $0.66 \times 10^{-3}T^2$ and $-1.02 \times 10^5 T^{-1}$, ΔG_e may be simply expressed in the form :

$$\Delta G_e = -87685 + 118.01 \, T - 9.65T \ln T \quad \text{relating to the reaction}$$

$$2/5 \; Nb + Cl_{2(g)} \rightleftarrows 2/5 \; NbCl_{5\,(v)}$$

1.1.2
Sulphidation Behaviour of Preoxidised Inconel 600 and Nimonic PE11 Alloys

H. L. Du,[1] D. B. Lewis,[2] J. S. Gray,[1] and P. K. Datta[1]

[1] SURFACE ENGINEERING RESEARCH GROUP, UNIVERSITY OF NORTHUMBRIA AT NEWCASTLE, NEWCASTLE UPON TYNE NE1 8ST, UK

[2] MATERIALS RESEARCH INSTITUTE, SHEFFIELD HALLAM UNIVERSITY, SHEFFIELD S1 1WB, UK

1 INTRODUCTION

The commercially available iron-, nickel-, and cobalt-based alloys rely for environmental protection on the selective oxidation of chromium and/or aluminium to form a dense, compact and adherent exterior oxide scale. Such conventional alloys were developed primarily for operation under highly oxidising conditions. However, there are several technological processes, e.g. coal gasification and petrochemistry, whose environments are generally characterised by low oxygen activity but high sulphur activity, these alloys are unlikely to form a protective oxide scale at the alloy/gas interface and unacceptably high rates of material degradation are observed. One approach is to preoxidise the components in order to form a significant coating of oxide and rely on this scale for protection. Some improvements in the degradation resistance of the alloys to sulphur-containing gases have been achieved by this method[1-7].

More recently, Du[8] and Datta et al[9] have demonstrated that preoxidation improved sulphidation behaviour of the Co-based alloys containing refractory metals in an atmosphere of $pS_2 \sim 10^{-1}$ Pa and $pO_2 \sim 10^{-18}$ Pa at 750°C. In particular, preoxidation greatly enhanced the sulphidation resistance of the Nb-containing alloys. The parabolic rate constants for the preoxidised specimens were four orders of magnitude smaller than those of the non-preoxidised samples. The enhancement of sulphidation resistance was considered to be caused by the preformation of a continuous α-Al_2O_3 scale. Even after 240 hours sulphidation, the α-Al_2O_3 scale was still effective. It was also shown that the preformed α-Al_2O_3 scale was much more effective than the preformed α-Cr_2O_3 scale in promoting durability in this particular sulphidising environment.

In this paper, the results from the sulphidation behaviour of two commercial alloys, Inconel 600 and Nimonic PE11, which were preoxidised at both high and low oxygen partial pressures are presented.

Table 1 Composition of Inconel 600 and Nimonic PE11 alloys

Alloy	Elemental content, wt%								
	Si	Al	Ti	Cr	Mn	Mo	Fe	Co	Ni
Inconel 600	0.3	-	0.2	16.0	0.3	-	9.5	0.5	bal.
Nimonic PE11	0.3	0.6	2.2	18.5	0.2	4.3	35.0	0.6	bal.

2   EXPERIMENTAL

The two commercial alloys, Inconel 600 and Nimonic PE11, were supplied by Inco Alloys Ltd and their compositions are listed in Table 1. The coupons, 15x10x1 (or 1.5) mm with 1 mm diameter hole drilled near a top short edge, were machined from the alloy sheets. The specimen surface was ground on emery paper to 1200 grit, following by ultrasonic cleaning and degreasing in acetone.

The prepared samples were preoxidised at $900^{O}C$ in open air ($pO_2 \sim 0.2 \times 10^5$ Pa) and in H_2/H_2O gas mixture ($pO_2 \sim 10^{-14}$ Pa). The sulphidation of the preoxidised and non-preoxidised specimens was carried out at $750^{O}C$ in $H_2/H_2S/H_2O$ gas mixture which was designed to produce an atmosphere of $pS_2 \sim 10^{-1}$ Pa and $pO_2 \sim 10^{-18}$ Pa at the test temperature. The sulphidation kinetics were achieved by means of a discontinuous gravimetric method. The preoxidised and sulphidised samples were examined using scanning electron microscopy (SEM), and analysed by energy-dispersive X-ray analysis (EDX). The phases presented were characterised by X-ray diffraction (XRD).

3   RESULTS

Sulphidation Kinetics

Figure 1 reveals the sulphidation kinetics of the preoxidised and non-preoxidised Inconel 600 and Nimonic PE11 alloys. The non-preoxidised Inconel 600 alloy sulphidised extremely rapidly during the first 2 hours of exposure and then the weight gains remained unchanged, which indicated that the substrate had been consumed. Indeed metallographic observations confirmed the complete consumption of the substrate. However, enhanced sulphidation resistance was observed on the preoxidised specimens. The samples pre-oxidised at high pO_2 sulphidised very slowly whilst the samples preoxidised at low pO_2 showed no significant weight changes during the first 2 hours of exposure. Up to 24 hours exposure, the samples preoxidised at low pO_2 showed superior sulphidation resistance to those preoxidised at high pO_2. After longer exposure, the samples preoxidised at low pO_2 sulphidised more rapidly than those at high pO_2. However, in both cases, the preformed oxide scales were, to some extent, still protective even after 48 hours

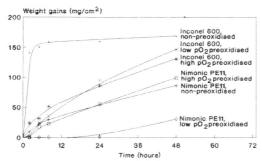

<u>Figure 1</u> Sulphidation kinetics of preoxidised and non-
preoxidised Inconel 600 and Nimonic PE11 alloys
sulphidation.

A remarkable improvement in sulphidation behaviour was
observed for the preoxidised Nimonic PE11 alloy. Although
large weight gains were recorded for the non-preoxidised
samples during the first 2 hours of sulphidation, little
weight changes were noticed on those preoxidised at both
high and low oxygen partial pressures. The immunity of the
samples preoxidised at low pO_2 to sulphidation was
persistent until 24 hours exposure and the preformed oxide
scales still provided protection even after the end of the
experiment (48 hours). However, preoxidation at high pO_2
offered only limited protection after prolonging
sulphidation. After 8 hours sulphidation, preoxidation
became detrimental and more rapid degradation was observed
on the samples preoxidised at high pO_2. Also, the
compositions of the substrates significantly influenced the
protectivity of the preformed oxide scales and particularly
in the case of low pO_2 preoxidation, the preformed oxide
scales on Nimonic PE11 alloy appeared more effective than
those on Inconel 600 alloy.

Morphologies

Morphologies of the preoxidised Inconel 600 and Nimonic
PE11 alloys at high oxygen partial pressure were quite
different from those at low oxygen partial pressure, as
shown in Figure 2. Preoxidation at high pO_2 brought about
the formation of Fe_3O_4 on both alloys. A Cr_2O_3 film formed
beneath the outer Fe_3O_4 layer on the Inconel 600 alloy,
whilst a mixture of Cr_2O_3 and TiO_2 was recognised on the
Nimonic PE11 alloy. In the case of low pO_2 preoxidation,
very limited amounts of Fe_3O_4 could be identified on both
alloys, and a pure Cr_2O_3 scale was evident on the Inconel
600 alloy, and a mixture of Cr_2O_3 and TiO_2 formed on the
Nimonic PE11 alloy. Moreover, more compact and finer-grained
oxide scales were observed on both alloys preoxidised at low
pO_2 than on the same alloys preoxidised at high pO_2.

Figure 3 depicts the sulphidation topographies of the
preoxidised and non-preoxidised Inconel 600 alloy after
various exposure periods. The non-preoxidised Inconel 600

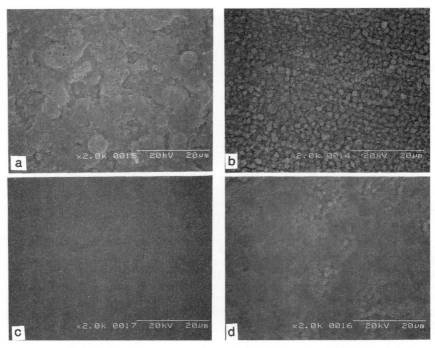

<u>Figure 2</u> Morphologies of preoxidised Inconel 600 and Nimonic
         PE11 alloys before sulphidation. (a) Inconel 600,
         high pO_2; (b) Nimonic PE11, high pO_2; (c) Inconel
         600 low pO_2; and (d) Nimonic PE11, low pO_2

alloy exhibited very rough and porous morphologies even
after only 30 minutes sulphidation, as shown in Figure 3
(a), whilst the preoxidised samples at high pO_2 suffered
only light sulphidation after 2 hours exposure and the
sulphide scales completely covered the sample surface
(Figure 3 (b)). The preoxidised samples at low pO_2 remained
unaffected. After prolonging sulphidation periods, the
preoxidised samples were severely attacked and the sulphide
scales developed became porous and cracked, as revealed in
Figure 3 (c); and observed to be similar to those on the
non-preoxidised samples. The XRD results indicated that
Ni_3S_2, Ni_4S_3 and $FeCr_2S_4$ were present on all the samples.

In the case of Nimonic PE11 alloy, preoxidation showed
more beneficial effect on improving sulphidation resistance
than in the Inconel 600 alloy. At the initial stages of
sulphidation (e.g. 2 hours), the preoxidised samples were
not attacked. The non-preoxidised samples, however, showed
severe corrosion, which was dominated by the formation of
iron and nickel sulphides, see Figure 4 (a). After 5 hours
sulphidation, the preoxidised samples at high pO_2 revealed
similar morphologies to the non-preoxidised samples, as
presented in Figure 4 (b), whilst preoxidised samples at low
pO_2 still remained intact. Even until 24 hours exposure, the

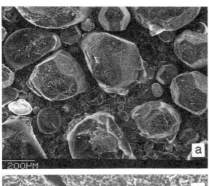
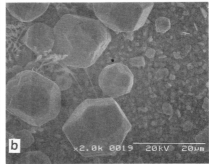
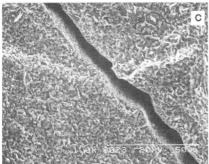

Figure 3 Morphologies of preoxidised and non-preoxidised
Inconel 600 alloy after
sulphidation. (a) non-
preoxidised, 30 minutes; (b)
high pO_2 preoxidised, 2 hours;
and (c) low pO_2 preoxidised,
5 hours

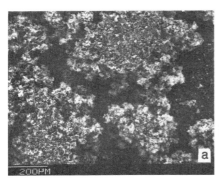
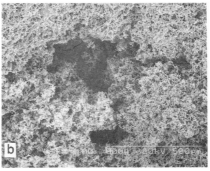

Figure 4 Morphologies of preoxidised and non-preoxidised
Nimonic PE11 alloy after
sulphidation. (a) non-
preoxidised, 2 hours; (b)
high pO_2 preoxidised, 5
hours; and (c) low pO_2
preoxidised, 24 hours

preoxidised samples at low pO_2 showed little evidence of corrosion, see Figure 4 (c), a unique scale formed on the surface was identified as Cr_2S_3 by XRD, but no iron and nickel sulphides were found.

Compositions

The most distinguishing feature between preoxidation carried out at high pO_2 and low pO_2 lay in the formation of Fe_3O_4, which was identified by XRD, on both Inconel 600 and Nimonic PE11 alloys. The oxide scales on Inconel 600 alloy preoxidised at high pO_2 consisted of Fe_3O_4 which was above a continuous Cr_2O_3 film, as shown in Figure 5. However no Fe_3O_4 was identified on the samples preoxidised at low pO_2. For Nimonic PE11 alloys, Cr_2O_3, TiO_2 and Fe_3O_4 characterised the preformed oxide scales at high pO_2 preoxidation, whilst no Fe_3O_4 was found on the samples preoxidised at low pO_2, as shown in Figure 6. Additionally, Al_2O_3 developed at the grain boundaries, which effectively blocked diffusion of sulphur and base elements during subsequent sulphidation.

The cross-section composition profiles for the preoxidised and non-preoxidised Inconel 600 alloy were not very different. Generally, the sulphide scales comprised two subscale layers. The outer-layer consisted of Ni_3S_2 and sulpho-spinel $FeCr_2S_4$. With increasing exposure periods, a small amount of Ni_3S_4 appeared in the outer-layer. The inner-layer mainly contained Cr_2S_3. However, a continuous Cr_2O_3 film was present in the inner-layer of Cr_2S_3, as shown in Figure 7 where the partial inner-layer near the substrate is presented. Also a band of Cr_2S_3 was found within the outer-layer after prolonged sulphidation, which may indicate the inner-layer of Cr_2S_3 grew in a discontinuous manner. Double layered sulphide scales also developed on the Nimonic

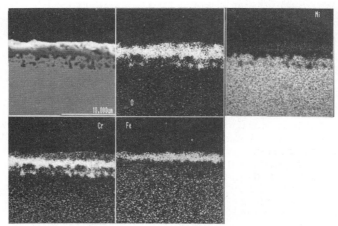

Figure 5 Electron image and Digimaps showing cross-sectioned composition of high pO_2 preoxidised Inconel 600 alloy before sulphidation

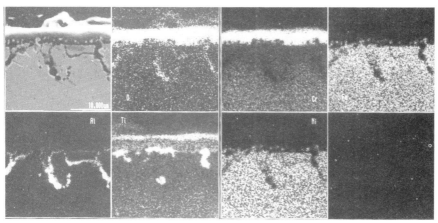

Figure 6 Electron image and Digimaps showing cross-sectioned composition profiles of low pO_2 preoxidised Nimonic PE11 alloy before sulphidation

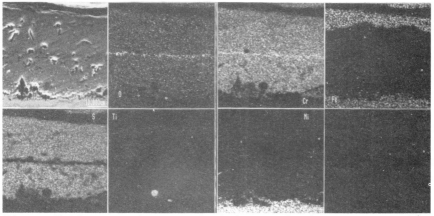

Figure 7 Electron image and Digimaps showing cross-sectioned composition profiles of high pO_2 preoxidised Inconel 600 alloy after 5 hours sulphidation

PE11 alloy. At the early stages of sulphidation (one hour), a major phase of pentlandite $(Ni,Fe)_9S_8$ and a minor amount of Ni_3S_2 with a sulpho-spinel $(FeCr_2S_4)$ characterised the outer-layer. The inner-layer consisted of titanium and aluminium sulphides. However with increasing sulphidation exposure, the pentlandite was transformed into Ni_3S_4 and a small quantity of NiS and $FeCr_2S_4$. However, the pentlandite phase was not identified on the preoxidised samples subjected to sulphidation. In contrast, a Cr_2S_3 scale was recognised over the preformed oxide scale, as showed in Figure 8, before the samples suffered severe corrosion. But the preformed Cr_2O_3 film disappeared after long term sulphidation (e.g. 48 hours), whilst some Al_2O_3, which still provided some protection, was identified within the inner-

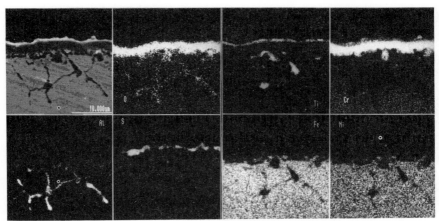

<u>Figure 8</u> Electron image and Digimaps showing cross-sectioned
composition profiles low pO_2 preoxidised Nimonic
PE11 alloy after 24 hours sulphidation

layer of Cr_2S_3.

Finally, it should be pointed out that low pO_2
preoxidation did not change the compositions after
sulphidation, but it did delay the onset of severe
corrosion.

4   DISCUSSION

The kinetics results indicated that preoxidation,
particularly at low oxygen partial pressure, significantly
enhanced the sulphidation resistance of both Inconel 600 and
Nimonic PE11 alloys. The improved sulphidation behaviour was
mainly attributed to the formation of the protective Cr_2O_3
scales which acted as an effective diffusion barrier,
separating the reactants, the metals and sulphur species.
For the Nimonic PE11 alloy, the discontinuous TiO_2 and Al_2O_3
subscales also effectively contributed to the enhanced
sulphidation resistance.

Two subscale layers characterised the sulphide scales
formed on all the specimens irrespective of the alloy
compositions and whether preoxidised or not. The outer-
layers mainly comprised Ni sulphides whilst the inner-layers
contained Cr_2S_3. It was noted that the preformed Cr_2O_3 film
was situated within the inner-layer of Cr_2S_3 near to the
Cr_2S_3/substrate interface. If the preformed Cr_2O_3 is assumed
as a natural marker, then the formation of both subscale
layers being mainly controlled by egress of base elements,
is indicated. Indeed, inward diffusion of sulphur is several
orders of magnitude slower than the outward migration of
nickel, iron and chromium[10] through a Cr_2O_3 scale. However,
considerable micro-imperfections or short circuit diffusion
paths existed within the oxide scales, formed during the
preoxidation period and subsequent cooling and handling; as

a result, the sulphur species were transported through these physical defects in the preformed oxide scale and approached the substrates to produce the Cr_2S_3 beneath the oxide.

The efficacy of preoxidation depends on the type of oxide formed and the perfection of the oxide film. It has been reported that Cr_2O_3 is a metal deficit semiconductor[11]. Apparently, preoxidation at low oxygen partial pressure would cause less non-stoichiometric defects in the Cr_2O_3 than preoxidation at high oxygen partial pressure. In the present study, the specimens preoxidised at low pO_2 offered enhanced sulphidation resistance. However, this is not the case for TiO_2 in the preoxidised Nimonic PE11 alloy. TiO_2 is known as an anion-deficit semiconductor; as a result, decreasing pO_2 would increase vacancies in TiO_2. Nevertheless, in the present situation, Cr_2O_3 is a dominant oxide and therefore low pO_2 caused a positive effect on the sulphidation resistance of Nimonic PE11 alloy. Meanwhile, the Fe_3O_4 developed on both alloys at high pO_2 acted as the diffusion paths for both anions and cations and consequently diminished sulphidation resistance. Furthermore, low pO_2 preoxidation favoured the formation of Cr_2O_3 rather than Fe_3O_4 and therefore, thicker Cr_2O_3 scales were produced during the same exposure period for low pO_2 preoxidation than for high pO_2 preoxidation; hence this would increase the length of immunity.

5   SUMMARY

Preoxidation significantly increased the sulphidation resistance of Inconel 600 and Nimonic PE11 alloys. Low pO_2 preoxidation showed more beneficial effect than high pO_2 preoxidation. The enhanced sulphidation behaviour of the alloys was attributable to the pre-formation of a Cr_2O_3 scale. Low pO_2 preoxidation suppressed the development of Fe_3O_4 and produced a thicker and less-defective Cr_2O_3 film and as a result provided a longer period of immunity. Also the preformed Cr_2O_3 scale on Nimonic PE11 alloy was more effective than that on Inconel 600 alloy at suppressing sulphidation.

REFERENCES

1. F.H. Stott, F.M.F. Chong and C.A. Stirling, "Conf. Proc. of Met. Soc. AIME/Met. Sci. Division ASM Symp. on High Temperature Corrosion in Energy Systems", ed. M.F. Rothman, Michigan, 1984.
2. D.B. Rao, K.T. Jacob and H.G. Nelson, <u>Metall. Trans.</u>, 1983, <u>14A</u>, 295.
3. S. Mrowec and M. Wedrychowska, <u>Oxid. Met.</u>, 1979, <u>13</u>, 481.
4. P.J. Hunt and K.N. Strafford, <u>J. Electrochem. Soc.</u>, 1981, <u>128</u>, 252.
5. C. Bresseleers et al, "Physical Sciences Commission of the European Communities Report EUR 6203 EN JRC", Petten (N.H.), 1979.
6. J.B. Wagner Jr., "Defect and Transport in Oxides", ed.

M.S. Smeltzer and Jaffee, 1974, p. 283.
7.   F.H. Stott and M.F. Chong, "Corrosion Resistant Materials for Coal Conversion Systems", ed. D.B. Meadowcroft and M.I. Manning, 1983, p. 491.
8.   H.L. Du, PhD Thesis (CNAA), Newcastle Polytechnic, 1991.
9.   P.K. Datta, K.N. Strafford, H.L. Du, B. Lewis and J.S. Gray, "Heat-Resistant Materials", Proc. of 1st Int. Conf., ed. K. Natesan, Fontana, Wisconsin, 1991, p. 323.
10. R.E. Lobnig, H.P. Schmidt, K. Hennesen and H.J. Grabke, Oxid. Met., 1992, 37, 81.
11. P. Kofstad, "High Temperature Corrosion", Elsevier Applied Science, London, 1988, Chapter 4, p.114.

1.1.3
Evaluation of Corrosion Resistance of Al_2O_3 and Cr_2O_3 Scales against a Chlorine-containing Environment

H. Chu,[1] P. K. Datta,[1] J. S. Gray,[1] and K. N. Strafford[2]

[1] SURFACE ENGINEERING RESEARCH GROUP, UNIVERSITY OF NORTHUMBRIA AT NEWCASTLE, NEWCASTLE UPON TYNE NE1 8ST, UK

[2] DEPARTMENT OF METALLURGY, UNIVERSITY OF SOUTH AUSTRALIA, THE LEVELS, SOUTH AUSTRALIA, 5095

1   INTRODUCTION

The philosophy of alloy design for high temperature application is essentially based on the selective oxidation of chromium or aluminium to form a compact and adherent Cr_2O_3 or Al_2O_3 scale which protects the underlying alloy substrate from environmental attack. The mechanisms describing the oxidation of such alloys in air or oxygen are reasonably well understood. In many industrial environments, however, materials may be exposed to mixed atmospheres containing various species, such as oxygen, sulphur and chlorine. Oxidation may remain to be the primary mode of attack in most oxygen-rich atmospheres, but this process can be substantially modified by the presence of small but finite amounts of sulphur, chlorine and carbon-bearing species particularly in an environment containing a low oxygen potential which is the characteristic of coal gasification.

The corrosion response of a series of HT alloys in various chlorine-containing environments has been reported[1-3]. One nickel-based, alumina-forming alloy, Haynes alloy 214, has been frequently investigated, and usually exhibited the best corrosion resistance amongst the commercial alloys tested, which included Inconel 601 and Hastelloy alloy X. The environmental durability of Haynes alloy 214 was usually attributed to the formation of an external Al_2O_3 scale. For chromia-forming alloys, considerable corrosion damage, together with substantial internal attack, and severe weight losses due to the formation and evaporation of chloride species, were observed.

In general, however, the corrosion behaviour of high temperature alloys in various chlorine-containing environments needs to be further investigated. It is particularly important to characterise the protectivity of both Cr_2O_3 and Al_2O_3 scales in such environments in order to design high temperature chloridation resistance alloys or coatings. The main aim of this study is therefore to evaluate the performance of Cr_2O_3 and Al_2O_3 scales, which are preformed on several Fe(Ni)CrAlX (X = e.g. Zr, Hf) type of alloys, in an environment of low oxygen potential and relatively high chlorine activity ($pO_2 \sim 10^{-15}$ Pa and $pCl_2 \sim 10^{-5}$ Pa) at 800°C - a combination typically encountered in coal gasification atmospheres.

2 EXPERIMENTAL

Eight model alloys have been selected as testing materials which, according to their chemical compositions (see Table 1), are classified as Al_2O_3-forming and Cr_2O_3-forming alloys. The alloys were vacuum induction melted and cast into cylindrical ingots of about 37 mm diameter. The outer layers (~ 2mm) of the ingots were machined off prior to coupon preparation. Rectangular specimens (10 x 8 x 2 mm) were then cut from these ingots, polished to 1200 mesh emery paper and subsequently degreased with acetone.

In order to form an external and compact oxide scale on each model alloy, preoxidation treatments were carried out for these testing materials in air at 800°C or 1000°C for 50 hours. The preoxidised specimens were then suspended on a silica sample holder which was placed in the hot zone of a horizontal reaction tube furnace and exposed to an $H_2/HCl/H_2O$ gas mixture designed to develop an environment comprising $pCl_2 \sim 10^{-5}$ Pa and $pO_2 \sim 10^{-15}$ Pa at 800°C for periods up to 16 hours. The details of the apparatus were described somewhere else[4]. The kinetics were determined by a discontinuous weight change method and the post-exposure morphological and compositional analyses of the scales were facilitated by SEM, EDAX and XRDA.

3 RESULTS

The corrosion kinetics results for preoxidised Al_2O_3-forming alloys following up to 16 hours exposure in the chlorine-containing environment at 800°C are shown in Figure 1. It is apparent that preoxidised FeCr8AlX type alloys exhibited excellent corrosion resistance in this environment

Table 1 Chemical composition of model alloys

	Alloy	Fe	Ni	Cr	Al	Zr	Hf	other
			Elemental content (wt%)					
	alloy 20	base		15.04	8.15	1.41		
Al_2O_3-	alloy 22	base		12.40	7.70		0.50	Ta=3.1
former	alloy 25M	base		12.40	7.80		2.80	
	alloy 26M	base		12.20	7.90		0.80	
	alloy 5	base	10.0	25.00				
Cr_2O_3-	alloy 14	0.14	base	24.75	2.41			
former	alloy 19	base		23.40	3.06	0.91		
	alloy 35	base	3.60	25.30			0.29	Si=0.2

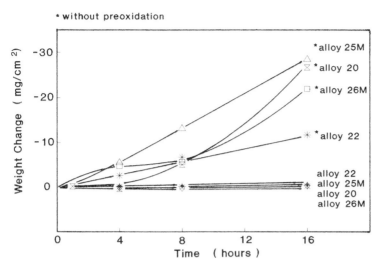

Figure 1 Corrosion kinetics of preoxidised FeCr8AlX-type
alloys subjected to $pCl_2 \sim 10^{-5}$ Pa and $pO_2 \sim 10^{-15}$ Pa
at 800^0C

and the relevant weight changes for four alloys tested were
all below -0.5mg/cm². In comparison, the as-received alloys
were severely attacked in the same environment and suffered
substantial weight losses, indicating that a preformed Al_2O_3
scale was an effective barrier against the attack of
chlorine-containing species.

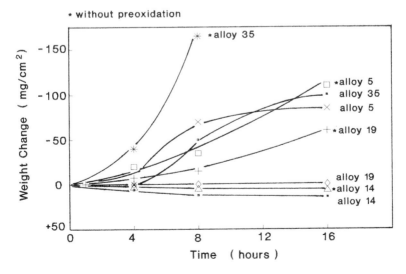

Figure 2 Corrosion kinetics of preoxidised Fe(Ni)25CrAlX-
type alloys subjected to $pCl_2 \sim 10^{-5}$ Pa and $pO_2 \sim$
10^{-15} Pa at 800^0C

Figure 2 represents the kinetic results for Cr_2O_3-forming alloys. The kinetic curves indicate that preoxidised alloy 5 and 35 were slightly attacked in the initial stage, but breakaway corrosion was observed after between 4 and 8 hours exposure to this environment and the weight changes respectively approached $-82mg/cm^2$ and $-115mg/cm^2$ after 16 hours testing. In comparison, preoxidised alloys 14 and 19 showed better corrosion resistance than alloys 5 and 35. When compared with as-received alloys, it is observed that preformed Cr_2O_3 scale improved the corrosion resistance of alloys 5 and 35 in short-term exposure of up to 4 hours, but the protection of Cr_2O_3 scale disappeared afterward. For alloy 19, however, preoxidation treatment did exhibit a beneficial effect in increasing the chloridation resistance of the alloy although, in general, a preformed Cr_2O_3 scale was not as an effective barrier as an Al_2O_3 scale against the attack of chlorine-containing species.

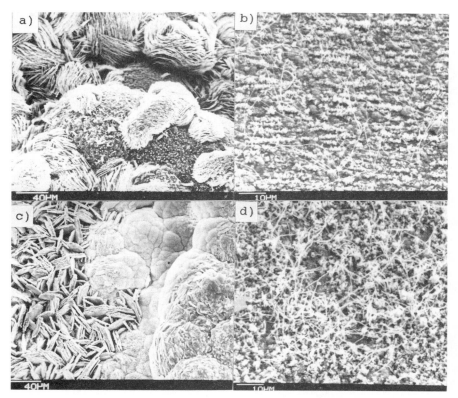

Figure 3 SEM micrographs of scales formed on alloys 25M and 26M in an oxychlorine environment ($pCl_2 \sim 10^{-5}$ Pa and $pO_2 \sim 10^{-15}$ Pa) at 800^0C. (a) Alloy 25M, as-received, 16h; (b) Alloy 25M, preoxidised, 16h; (c) Alloy 26M, as-received, 16h; (d) preoxidised, 16h

Representive scale morphologies for Al$_2$O$_3$-forming alloys are depicted in Figure 3. Clearly, the Al$_2$O$_3$ scales preformed on alloys 25M and 26M remained to be effective in preventing the alloy substrates from chloridation after up to 16 hours exposure. However, as-received materials, which had no protection of Al$_2$O$_3$ scales, underwent serious corrosion damage as indicated by the formation of huge nodulous corrosion products emerged from the grain boundaries in the scale. Figure 4 shows the scale micrographs of Cr$_2$O$_3$-forming alloys 5 and 35 following exposure in the same environment. It is apparent from Figure 4 that both (as-received) alloys were severely attacked and a great amount of corrosion products was formed on the alloy surface after corrosion tests. As indicated in Figure 4 (b and d), preoxidation treatment had little effect in improving the chloridation resistance of the alloys, suggesting that preformed Cr$_2$O$_3$ scale was not an ideal barrier scale which could protect the alloy substrate effectively from the attack of chlorine-bearing species.

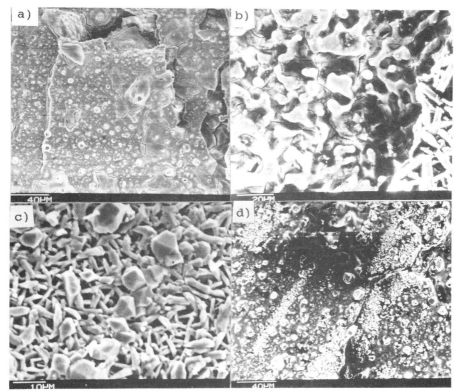

Figure 4 SEM micrographs of scales formed on alloys 5 and 35 in an oxychlorine environment (pCl$_2$ ~ 10^{-3} Pa and pO$_2$ ~ 10^{-15} Pa) at 800^{0}C. (a) Alloy 5, as received, 4h; (b) Alloy 5, preoxidised, 16h; (c) Alloy 35, as-received, 4h; (d) preoxidised, 16h

Table 2 Phases (main and secondary) of scales formed on
preoxidised alloys before and after chloridation
tests

Alloy type	As-preoxidised	Post-chloridation
Al_2O_3-former	M: Al_2O_3 S: Cr_2O_3, ZrO_2 or 　　HfO_2	M: Al_2O_3 S: Cr_2O_3, $CrCl_2$, 　　ZrO_2 or HfO_2
Cr_2O_3-former	M: Cr_2O_3 S: Al_2O_3, ZrO_2 or 　　HfO_2	M: $CrCl_2$, Cr_2O_3 S: Al_2O_3, $FeCl_2$, 　　ZrO_2 or HfO_2

Table 2 summarises the main phases of scales formed on
preoxidised Al_2O_3- and Cr_2O_3-forming alloys before and after
chloridation. For Al_2O_3-forming alloys, preoxidised scale
consists of mainly Al_2O_3 with some ZrO_2 or HfO_2. After
chloridation, Al_2O_3 remains to be the main phase of the scale
although small amount of $CrCl_2$ is also detected in the scale
particularly after long-term corrosion tests. In the case
of Cr_2O_3-forming alloys, whilst Cr_2O_3 is the main phase of the
oxide scale before chloridation tests, both $CrCl_2$ and Cr_2O_3
are the dominating corrosion products following exposure to
the chlorine-containing environment.

Figure 5 displays the elemental distribution maps of
a cross-section of preoxidised alloy 35 following 16 hours
chloridation. The results indicate that the scale formed was
divided into two layers. The outer layer was abundant in Cl
and Cr but lean in Fe, whilst the inner layer, which seemed
to be porous, contained Fe, Cr and Cl.

4　DISCUSSION

The experimental results have indicated that
preoxidation treatment significantly enhanced the corrosion
resistance of FeCr8AlX type alloys in an environment with
$pCl_2 \sim 10^{-5}$ Pa and $pO_2 \sim 10^{-15}$ Pa at 800°C as shown in Figure
1. The relevant weight changes for the preoxidised alloys
were very small compared with those of as-received alloys;
for example, preoxidised alloy 25M experienced a weight
change of about -0.5mg/cm² after 16 hours exposure, whilst
as-received alloy 25M suffered a weight change of -28mg/cm².

Examination of the scales for preoxidised FeCr8AlX type
alloys following chloridation tests confirmed that preformed
Al_2O_3 scale was retained with small amounts of ZrO_2, HfO_2 and
Cr_2O_3 on the alloys after exposure. No corrosion products (
$FeCl_2$ and $CrCl_2$) were visibly formed on the reaction tube
after the corrosion tests although a trace amount of $CrCl_2$
was identified in the scales of preoxidised alloys 20 and 22
after prolonged exposure (16 hours).

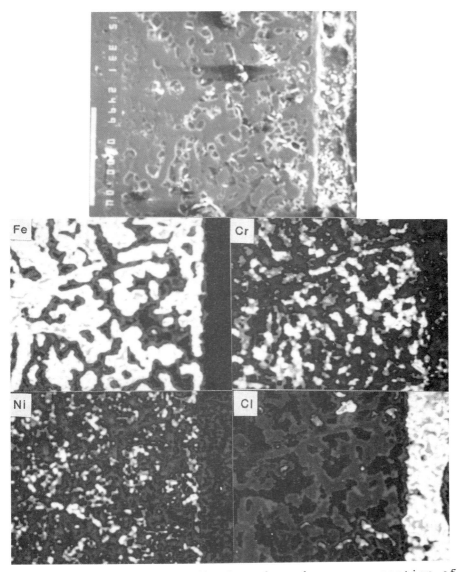

<u>Figure 5</u> Elemental distribution through a cross-section of preoxidised Alloy 35 after 16 h exposure in an oxychlorine environment ($pCl_2 \sim 10^{-5}$ Pa and $pO_2 \sim 10^{-15}$ Pa) at 800^0C

These results clearly indicated that preoxidation treatment was effective in suppressing chloridaton attack on FeCr8AlX type alloys. Because of the formation of an external Al₂O₃ scales on the alloys, the alloy substrates were separated from exposure to the chlorine-containing environment, thereby preventing the formation of FeCl₂ by the following reaction:

$$Fe(s) + Cl_2(g) = FeCl_2(l) \tag{1}$$

The chloridation behaviour of four preoxidised Fe(Ni)25CrAlX type alloys (i.e. alloys 5, 14, 19 and 35) in the same environment ($pCl_2 \sim 10^{-5}$ Pa, $pO_2 \sim 10^{-15}$ Pa) at 800°C indicated that the protectivity of the preformed Cr_2O_3 scale was not as protective as an Al_2O_3 scale, particularly after prolonged exposure (>4 hours). Breakaway corrosion was observed on alloys 5 and 35, both containing no aluminium, after 4-8 hours corrosion test. However, such breakaway corrosion phenomenon was not noticed on the other alloys (14 and 19) which contained small amounts of aluminium (2.5%). Examination of the scales of the preoxidised alloys confirmed that the preformed Cr_2O_3 scale usually remained on the alloy following short term (< 4 hours) exposure in the oxychlorine environment. However, $CrCl_2$ gradually became a predominant phase present in the scales after longer term corrosion testing, particularly for alloys 5 and 35. Furthermore, the deposition of corrosion products on the reaction tube proved to be essentially $FeCl_2$ with some $CrCl_2$, indicating that the evaporation of $FeCl_2$ and some $CrCl_2$ was mainly responsible for the weight losses observed on the iron-based alloys.

According to the relevant thermodynamic data[5], Cr_2O_3 scale is stable and would be retained on the alloy surface in the present oxychlorine environment. Thus, the formation of an external Cr_2O_3 scale on the Fe25CrAlX type alloys by preoxidation treatment prohibited the formation of $FeCl_2$ via reaction (1) during the initial stage of exposure. However, it appears that, with lengthening exposure, chlorine-containing species migrated through the Cr_2O_3 scale and reached the alloy/oxide interface although the relevant transport route was not clear. There are two ways in which chlorine-containing species may be transported through a oxide scale; i.e. dissolution/diffusion and gas penetration mechanisms. Due to the scarcity of solution and diffusion data for chlorine in the oxides including Cr_2O_3 and Al_2O_3, it is difficult to examine the migration mechanism of chlorine species within the Cr_2O_3 or Al_2O_3 scales. Nevertheless, it seems that chlorine-containing species were able to penetrate through a Cr_2O_3 scale much faster than an Al_2O_3 scale under the same experimental conditions although the mechanism in which chlorine species were transported remained to be understood.

When the partial pressure of chlorine at the oxide/alloy interface was established at a level higher than the dissociation pressure of the chlorides, the relevant chlorides, including $FeCl_2$ and $CrCl_2$, were thermodynamically able to form. Chloridation of iron firstly occurred at the alloy/oxide interface in the initial stage of exposure in the present oxychlorine environment. With the depletion of iron in the substrate and further establishment of partial pressure of chlorine in the long-term corrosion tests, chromium also tended to be chloridised. There are two possible ways in which $CrCl_2$ may be formed at the alloy/oxide interface. One is the direct chloridation of chromium with chlorine species via the reaction

$$Cr(s) + Cl_2(g) = CrCl_2(s) \tag{2}$$

The other possible way to form $CrCl_2$ is through the transformation of Cr_2O_3, i.e.

$$Cr_2O_3(s) + 2Cl_2(g) = 2CrCl_2(s) + 1.5O_2(g) \tag{3}$$

The additional condition for the reaction (3) taking place is that the oxygen potential at the site is extremely low. Obviously, $CrCl_2$ is able to form at the alloy/oxide interface where the oxygen potential is possibly as low as the dissociation pressure of the oxide. As reaction (3) continues, the preformed Cr_2O_3 scale may finally breakdown, consequently resulting in the breakaway corrosion of the alloys as observed for alloys 5 and 35. However, the breakdown of the preformed Cr_2O_3 scale can be prevented by the existence of a small amount of aluminium in the alloys. If $CrCl_2$ was formed by reaction (3), a subscale of Al_2O_3 would then be developed through reaction between aluminium in the alloy substrate and the oxygen released by the reaction, thus greatly reducing the migration of chlorine-containing species within the scale and eliminating or limiting the consequent breakdown of the Cr_2O_3 scale as observed for the alloy 19.

5 CONCLUSIONS

1) Al_2O_3 scale formed on FeCr8AlX type alloys following preoxidation is effective in enhancing the chloridation resistance of the alloy.
2) The protectivity of the preformed Cr_2O_3 scale on Fe(Ni)25CrAlX type alloys against chlorine-containing environments is inferior to that of Al_2O_3 scale.
3) The corrosion damage of preoxidised Fe(Ni)CrAlX type alloys is mainly caused by the formation of $CrCl_2$ and $FeCl_2$ beneath the oxide scale.
4) Al_2O_3 scale is more impervious to chlorine species than Cr_2O_3 scale. However, the detailed diffusion mechanisms of chlorine in both type of scales need to be further investigated.

REFERENCES

1. F.H.Stott, R.Prestott, P.Elliott and M.H.J.M.Al'Atia, High Temperature Technology, 1988, 6, 115.
2. P.Elliott and G.Marsh, Corrosion Science, 1984, 24, 397.
3. J.M.Oh, M.J.McNallan, G.Y.Lai and M.F.Rothman, Metallurgical Transactions, 1986, 17A, 1087.
4. H.Chu, PhD Thesis, Newcastle Polytechnic, UK, 1991.
5. O.Kubaschewski and C.B.Alcock, "Metallurgical Thermochemistry", 5th Edition, Pergamon Press, 1977.

1.1.4
Microstructural Stability of Aluminide Coatings on Single Crystal Nickel-base Superalloys

W. F. Gale,[1] T. C. Totemeier,[1] A. J. Whitehead,[2] and J. E. King[1]

[1] DEPARTMENT OF MATERIALS SCIENCE AND METALLURGY, UNIVERSITY OF CAMBRIDGE, PEMBROKE STREET, CAMBRIDGE CB2 3QZ, UK

[2] MATERIALS DIVISION, DEPARTMENT OF MECHANICAL, MATERIALS AND MANUFACTURING ENGINEERING, HERSCHEL BUILDING, THE UNIVERSITY, NEWCASTLE UPON TYNE NE1 7RU, UK

1 INTRODUCTION

Improvement of the oxidation resistance of nickel-base superalloys can be achieved by the use of aluminide diffusion coating [1]. The high activity diffusion coating process involves the formation of a B2 type (nominally NiAl) coating by diffusion of aluminium into the superalloy and reaction with nickel from the superalloy. Although the use of this process is well established, the microstructures developed in these coatings and relationships with mechanical properties are not yet fully understood. The nature of the coating formed during aluminisation is strongly influenced by the composition of the substrate and hence observations tend to be specific to a particular alloy system. In the present investigation an examination employing transmission electron microscopy and mechanical testing has been made of a relatively simple single crystal substrate (containing principal alloying additions of Co, Cr, Al, Mo and V - for further details see Tawancy et al [2]) containing features common to a range of γ / γ ' superalloys employed as aero engine gas turbine materials. Attention is focussed on the changes occuring in the coating during post-coating heat treatment and subsequent exposure, as these have been found to be very marked and to have a strong influence on mechanical properties. To assist in the understanding of microstructural development in this system, an initial study of bulk aluminides with coating-like compositions has also been conducted. The use of bulk aluminides has the advantage that, unlike a coating, the overall composition of the alloy remains constant and hence the number of variables governing microstructural development is reduced. As a first stage, the extent to which the phases present in the coating can be duplicated is examined.

2 EXPERIMENTAL TECHNIQUES

Samples of the single crystal superalloy were initially solution treated for 4 hr at 1260 °C followed by gas-fan quenching to 800 °C and air cooling to room temperature. Aluminisation was performed for 4 hr at 870 °C followed by a diffusion treatment consisting of 1 hr at 1100 °C. The material was then aged for 16 hr at 870 °C. Throughout the coating and heat-treatment processes an argon atmosphere was employed. A range of coatings with thicknesses of between 10 and 50 µm could be produced in this manner depending upon the sample load employed. Subsequent exposure treatments at temperatures of between 850 and 1100 °C were performed using durations of up to 140 hr in air or a 10^{-4} mbar vacuum. Edge-on samples were prepared using the technique described previously [3] and examined using transmission electron microscopy (TEM). Supplementary studies were conducted using scanning electron microscopy (SEM) and both TEM and SEM based energy dispersive X-ray analysis (EDS). These results were compared with Vickers microhardness and nanoindentation studies. Coating fracture strains were determined using 3-point bending and tensile studies on thin walled specimens. Two model alloys were prepared by arc melting and were examined in the as-cast condition using TEM. Compositions of the two models corresponded to the coating centre region and to the portion of the coating adjacent to the substrate.

3 RESULTS AND DISCUSSION

Precipitation in the Coating and Adjacent Substrate.

Two main categories of microstructural changes were found to occur in the coating during post-coating heat treatment and subsequent exposure. These were, firstly, the formation of a number of phases derived directly from decomposition of the B2 coating matrix (commonly denoted as β phase) and, secondly, precipitation produced by rejection of ternary additions, such as chromium, from the B2 phase.

β Decomposition Products. Following coating and diffusion treatment, the coating was found to be nickel-rich throughout its thickness [3]. Under these conditions the precipitation of more highly nickel-rich phases, in particular the $L1_2$ type γ' phase (nominally $Ni_3(Al,Ti)$), is not surprising. However, it was noted that the manner in which precipitation occurred was heavily influenced by the heat treatment / exposure temperature employed. During low temperature (850 to 950 °C) heat treatment and exposure γ' formation took place predominantly within the body of the coating, rather than at its extremities. Under these conditions individual γ' precipitates were formed either directly from the β phase (on β - β grain boundaries and intragranularly) or on $M_{23}X_6$ phases precipitated during the coating / diffusion treatment. Further precipitation of $M_{23}X_6$ and γ', on the initial γ' particles, led to the formation of an intricate γ' / $M_{23}X_6$ network. Figure 1 shows typical γ' precipitates formed during low temperature (850 °C) treatment. The occurrence of these precipitates did not correlate with compositional changes in the coating induced by either oxidation of the coating surface or interdiffusion with the substrate. Instead, the formation of γ' could be accounted for by a simple ageing process following reductions in nickel solubility between the 1100 °C diffusion treatment temperature and subsequent lower temperature (850 - 950 °C) ageing and exposure. Indeed, on cycling back up to 1100 °C the precipitation process was found to be completely reversible.

In work on 850 to 950 °C exposure, a limited amount of preferential formation of γ', associated with loss of aluminium to the outer Al_2O_3 layer, was observed near the coating surface. In contrast, at 1100 °C, γ' formation at the coating surface and penetration down β - β grain boundaries (figure 2a) into the main body of the coating occurred extensively. Furthermore, at 850 - 950 °C, interdiffusion with the substrate was not found to have a marked influence on the microstructure of the coating within the time scale examined (up to 140 hr exposure). At 1100 °C the formation of a prominent continuous layer of γ' (figure 2b) associated with compositional changes in the coating induced by interdiffusion was observed on the coating side of the coating-substrate interface. The combined effect of oxidation and interdiffusion after 140 hr at 1100 °C was the formation of islands of β surrounded by γ' (with outer layers of γ / γ' in the case of relatively thin, i.e. 30 μm or less, coatings).

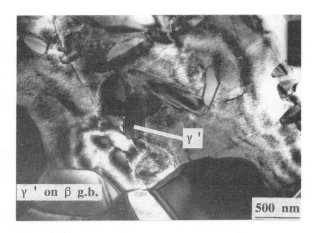

a: Intragranular and β grain boundary γ'.

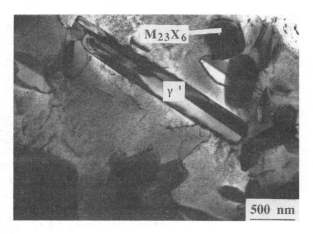

b: Intragranular γ' with γ' / $M_{23}X_6$ clusters.

Figure 1 γ' precipitated in a sample exposed for 1.5 hr at
         850°C in a 10^{-4} mbar vacuum

Although γ' (and γ) were the most marked β decomposition products, other
phases derived from β were also observed. In the titanium rich regions of the
coating (nearest to the substrate from which the titanium originated) the formation of
the L2$_1$ β' phase (nominal composition Ni_2AlTi) was observed, as it was in work

by Shen et. al. on aluminised René 80 [4]. This phase is a β derivative which involves the formation of separate titanium and aluminium sublattices in titanium rich β. At 1100 °C titanium enrichment was observed in the outer portion of the coating due to extensive loss of aluminium into the oxide layer, and this was found to have the effect of promoting the formation of a titanium rich phase Ni_4Ti_3. The production of two other phases derived from the β phase was also observed in this region and these will be discussed further in the section on bulk alloy coating analogues.

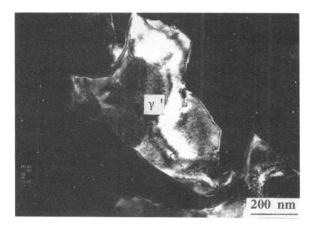

a: Formation of γ' along β - β grain boundaries (sample exposed for 16 hr in air).

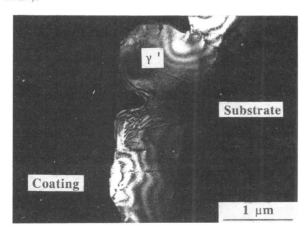

b: Preferential formation of γ' in the coating near the coating - substrate interface (sample exposed for 140 hr in a 10^{-4} mbar vacuum).

Figure 2 **γ'** precipitation in sample exposed at 1100^0C

Precipitation of Chromium Containing Phases. In addition to the formation of a range of β derivatives, the rejection of chromium from the β phase was found to be an important factor in coating microstructural development. In the as-diffusion treated state $M_{23}X_6$ precipitation was observed (for which M was predominantly chromium with some molybdenum and X may be presumed to be carbon). However, the supply of carbon is limited and this was found to favour the formation of other phases during 850 to 950 °C exposure. Initially α - Cr was formed extensively; this had the effect of decreasing the ratio of chromium to molybdenum in the coating and promoting the formation of molybdenum rich phases, principally σ (figure 3 shows σ precipitation on an MX phase incorporated from the substrate during coating formation) although χ and M_6X were also observed. A combination of chromium and titanium rejection simultaneously led to the formation of Cr_2Ti. At 1100 °C similar behaviour was observed, although re-solutioning of $M_{23}X_6$ occurred during holding at this temperature. In the substrate adjacent to the coating the compositional disturbance produced during the coating process (consisting of a reduction of the nickel to chromium / molybdenum ratio) resulted in extensive precipitation of $M_{23}X_6$ (850 - 950 °C treatment) and σ (850 - 1100 °C treatment).

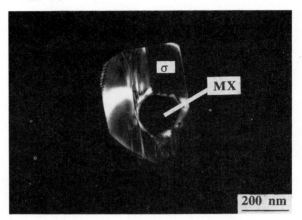

Figure 3 σ-phase precipitation on MX (sample exposed for 1.5 hr at 850^0C in air)

Bulk β - Phase Based Intermetallics as Analogues for Coating Microstructures.

In order to further understanding of microstructural development in coatings the production of bulk alloys which (unlike the coatings) have fixed overall compositions is of value. An effort has therefore been made to develop suitable model alloys, and initial progress with these materials in the as-cast condition is now reported. Consider first an alloy whose composition is based on that of the main body of the coating. The bulk alloy was found to consist of β dendrites containing extensive α - Cr precipitation, together with some $M_{23}X_6$. Interdendritically, σ-phase was extensively precipitated, although occasional pockets of γ ' (or in a few cases γ / γ ') could be observed. These initial observations suggest that the bulk alloy at least mimics the phases present in the coating, although further work on homogenised alloys is obviously required. Of greater interest was a bulk alloy intended to correspond to the portion of the coating adjacent to the substrate. In this case β dendrites containing extensive α - Cr precipitates were observed as before. However, in addition to these phases further microstructural features were observed within the β dendrites.

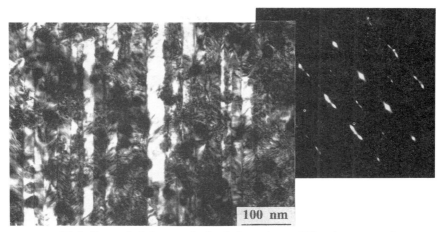

a: Main figure: lath features in the β phase, inset: diffraction pattern from
      this region showing the formation of diffuse scattering.

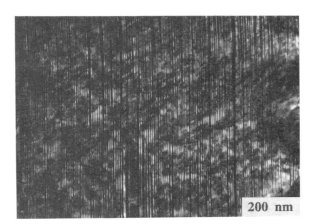

b: Twin like features observed in β.

Figure 4 β-phase derivative (as-cast model alloy for coating
        adjacent to substrate)

Diffuse lobes emanating from the B2 superlattice and fundamental reflections
could be observed in diffraction patterns (figure 4). Although the formation of
diffraction anomalies is familiar from studies of binary Ni-Al B2 phase, the contrast
anomalies were markedly different. The lobes were generally observed to originate
from prominent needle-like features within the β phase. For a given triplet
consisting of a B2 diffraction spot and two lobes it was observed that the B2 spot
illuminated the background β whilst each of the two lobes corresponded to different
needles. Within some β dendrites large regions (in some cases covering almost the
entire dendrite) gave rise to these lobes and each lobe corresponded to one variant of
twin-like striations in the β phase. Clearly, these features indicated the presence of a
B2 transformation product, although it would appear that since the B2 spots (which
were retained) and the lobes gave rise to different features within the β dendrites,
that the B2 and the new phase coexisted rather than the B2 having been entirely
replaced. These features are of considerable interest, because in coatings they were
observed simultaneously with another new superlattice also coexisting with the β

phase. In the coatings it appeared that the new phase was separate from that associated with the lobes around the B2 spots. Support for this suggestion was provided by the observation in the bulk alloy that the strong contrast associated with the lobes was not accompanied by the presence of the other new phase.

In the model alloy for the inner region of the coating, α - Cr was the only chromium based phase observed. Chromium became incorporated interdendritically where it stabilised the presence of the γ phase. As a result, interdendritic regions consisted of a γ ' coating around the β (containing numerous antiphase boundaries presumably as a result of the effect of chromium on order in this phase) which surrounded a superalloy-like region of γ ' in a γ matrix. In summary, although the bulk alloys do not yet directly resemble coating microstructures, it has been found that they can prove useful in understanding the apparently more complex precipitation behaviour in real coatings.

Implications of the Observed Precipitation Behaviour for Coating Mechanical Properties.

In order to examine the influence of the various microstructural features on mechanical properties, three distinct sample conditions were examined. These were: as-fully heat treated (as a base-line), 140 hr 850 °C air exposed (which resulted in extensive development of σ phase and some $M_{23}X_6$ in the substrate), and 140 hr 1100 °C air exposed (which resulted in widespread γ ' formation within the coating and the replacement of β - β grain boundaries with β - γ ' boundaries).

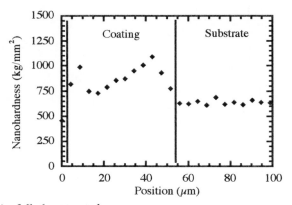

a: As-fully heat treated.

Figure 5 Nanohardness profiles after various post-coating heat treatments (Figure 5 continues on next page)

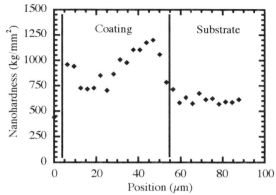

b: Exposed 140 hr 850 °C in air.

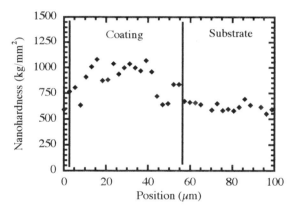

c: Exposed 140 hr 1100 °C in air.

Figure 5 Nanohardness profiles after various post-coating heat treatments

    The hardness profiles of the as-fully heat treated and the 850 °C exposed samples (figures 5a and b) were fairly similar. Coating hardness increased as the coating became more nickel rich; this occurred on traversing towards the substrate (i.e. away from the original source of aluminium during coating formation), and also in a region immediately next to the coating surface (where aluminium loss to the surface oxide layer depleted aluminium). This behaviour is in line with that of many intermetallics whose hardnesses are markedly increased by deviations from stoichiometry. Adjacent to the substrate (and, possibly, near the coating surface), where extensive $\gamma\,'$ formation occurred in the nickel rich β, a drop in hardness was observed. By comparison precipitation of chromium-bearing phases within the coating appeared to have relatively little effect, since hardness did not change markedly either at the interfaces between regions of extensive $M_{23}X_6$ formation and the rest of the coating or after 850 °C exposure (which greatly enhanced the formation of chromium containing phases).

    In the 1100 °C exposed sample (figure 5c), more variable results were obtained. The extensive formation of $\gamma\,'$ within the β matrix led to a considerable degree of scatter in the hardness values. Some softening occurred at the coating margins where large $M_{23}X_6$ precipitates were re-dissolved and where β was largely replaced by γ'. Of these effects, the γ' formation appeared to be most important as there was no strong correlation between the softened regions and those which had originally contained carbides.

Whilst the indentation studies described above showed no dramatic change between the various heat treatments, the situation with regard to coating fracture behaviour was found to be radically different. In the as-fully heat treated sample cracks nucleated at the coating surface and then propagated (roughly perpendicular to the coating surface) through the coating. The cracks generally followed β - β grain boundaries, although crack deflection and splitting at carbide phases was observed. Under these conditions room temperature coating fracture strains were in the range 0.52 - 0.65 %. The cracks formed in the coating were relatively blunt and did not penetrate significantly into the substrate. In the 140 hr 850 °C exposed sample, cracking extended into the substrate along σ phase needles aligned along $<110>_\gamma$ directions. Although the secondary cracks halted when substrate free of σ phase was reached the cracks remained sharp. The presence of these sharp cracks is of interest as they could provide effective initiation sites for fatigue crack growth, especially in samples subjected to monotonic preloading below the coating brittle - ductile transition temperature (around 750 °C).

The measured coating fracture strains for the 140 hr 850 °C exposed sample tested at room temperature (0.57 to 0.62 %) were similar to those in the as-fully heat treated sample. Although more extensive precipitation of chromium-containing phases occurred in the exposed sample, these did not substantially influence the behaviour of the already brittle coating. Some γ' was present on β - β grain boundaries (and intragranularly) in the lower portion of the coating; however, this was interspersed with brittle carbide phases and, similarly, did not have a discernible influence on crack propagation. In the 140 hr 1100 °C exposed sample much more extensive γ' precipitation occurred and $M_{23}X_6$ was re-dissolved. These changes resulted in greatly enhanced room temperature coating fracture strains in the range 5 - 8 %. In this case crack initiation occurred sub-surface, in the β phase. Cracks propagated through the β either transgranularly or along grain boundaries until they were stopped by meeting further γ'. Unlike β, polycrystalline γ' is not intrinsically brittle (although γ' - γ' interfaces can be weak [5]) and the presence of γ' was found to have a marked toughening effect, especially where coarse $M_{23}X_6$ precipitates have been removed.

4 CONCLUSIONS

An investigation of the stability of the microstructures of aluminide diffusion coatings during post-coating thermal exposure has been presented and correlated with coating mechanical properties. As a result of this study, the following conclusions have been drawn.

- The formation of γ' from an initially nickel rich coating represented the predominant microstructural change in the coating. During low temperature (850 - 950 °C) heat treatment and exposure, precipitation was dominated by a simple ageing process. However, at a higher temperature (1100 °C) interdiffusion with the substrate and oxidation of the coating surface were important. Loss of aluminium from the β phase also resulted in the formation of new superlattice phases within the coating. Simultaneously with the formation of γ', titanium rejection from the β phase resulted in the formation of β' and Ni_4Ti_3. In addition, extensive precipitation of chromium containing phases occurred both in the substrate and in the adjacent coating. The sequence of precipitation steps observed has been characterised and correlated with compositional changes in the coating.

- In micro- and nanohardness investigations the key factors in coating hardness were found to be deviations in stoichiometry of the β phase and γ' formation. The formation of chromium-containing phases had comparatively little effect on coating hardness.

- Coating fracture behaviour has been found to be markedly influenced by the microstructural changes. Extensive formation of γ' at 1100 °C increased the coating fracture strain from 0.52 - 0.65 % to 5 - 8 %. σ phase precipitation in the substrate produced sharp secondary cracks in the substrate below the relatively coarse cracks developed in the coating.

- Initial work on bulk alloys suggests that a number of coating microstructural features, such as σ phase, α - Cr, and β derivatives can be re-created in bulk samples.

ACKNOWLEDGEMENTS

The authors wish to express their thanks to Rolls-Royce plc and the Science and Engineering Research Council for financially supporting the project and W.F.G. The co-operation of Dr. T.N. Rhys-Jones and Mr. S. Williams of the Materials and Mechanical Research division of Rolls-Royce plc is gratefully acknowledged. Funding for the Nanoindenter II and A.J.W. was provided by the Materials Commission of the Science and Engineering Research Council. T.C.T. acknowledges the support of the Marshall Aid Commemoration Commission, whilst J.E.K. thanks British Gas plc and the Fellowship of Engineering for support. Laboratory facilities were provided by Professors C.J. Humphreys (Cambridge) and T.F. Page (Newcastle).

REFERENCES

1. T.N. Rhys-Jones in 'Materials Development in Turbo-Machinery Design' (Editors D.M.R. Taplin, J.F. Knott and M.H. Lewis), Institute of Metals, London and Parsons Press, Dublin 1989, p. 218.

2. H.M. Tawancy, N.M. Abbas and T.N. Rhys-Jones, Surf. Coat. Technol., 1991, 49, 1.

3. W.F. Gale and J.E. King, Metall. Trans. A, in press.

4. P. Shen, D. Gan and C.C. Lin, Mater. Sci. Eng., 1986, 78, 163.

5. M. Yamaguchi and Y. Umakoshi, Prog. Mater. Sci., 1990, 34, 1.

1.1.5
Development of Protective Coatings on Ti-6Al-4V by Laser Surface Alloying

A. Y. Fasasi,[1] S. K. Roy,[2] A. Galerie,[1] M. Pons,[1] and M. Caillet[1]

[1] LABORATOIRE SCEINCE DES SURFACES ET MATÉRIAUX CARBONÉS, E.N.S. D'ELECTROCHIMIE ET D'ELECTROMÉTALLURGIE DE GRENOBLE, INSTITUT NATIONAL POLYTECHNIQUE DE GRENOBLE, B.P. 75 – DOMAINE UNIVERSITAIRE, F-38402 SAINT-MARTIN D'HÈRES, FRANCE

[2] METALLURGICAL ENGINEERING DEPARTMENT, INDIAN INSTITUTE OF TECHNOLOGY, KHARAGPUR, 721302, INDIA

1 INTRODUCTION

For high temperature services, Ti-6Al-4V is an industrial alloy needing suitable protection. In order that the bulk properties of the alloy remain unchanged, this protection must be done in a way that only the surface region of the alloy is modified. In this way, the use of a laser or high energy beam for surface modification is one of the suitable techniques. The rapid heating and cooling associated with laser treatment result in a wide variety of desirable and controllable microstructures[1] which can be obtained by simply selecting the proper process parameters. In addition, laser-treated materials can be enriched with the properly chosen additive elements far in excess of their solubility limits.

The three major elements usually considered in this area are chromium, aluminium and silicon. The usefulness of chromium above 1000^0C is limited due to the formation of volatile CrO_3 [2]. Aluminium and silicon, on the other hand, lead to the formation of Al_2O_3 and SiO_2 that can effectively reduce the diffusion of oxygen and titanium, but the amount of these elements necessary to form a continuous oxide layer on titanium and its alloys is still a subject of controversy.

Several workers have tried to improve the mechanical properties and oxidation resistance of titanium[3-5] and

titanium aluminides, Ti_3Al and $TiAl$. These aluminides maintain good strength up to 1000^0C, but their use is limited above 650 and 1040^0C respectively[6]. Of particular importance among the works already done on Ti-Al systems is the paper by Welsch et al[7] who worked on different aluminium contents to reach the conclusion that a mixture of TiO_2 and Al_2O_3 external sublayer developed for aluminium contents greater than a critical value of 25at%. Meier et al[8] in their own work, gave a value of 64at%. However all the authors agreed that the higher the aluminium content, the higher the oxidation resistance.

On bulk Ti-Si alloys, concentrations as high as 10wt% had been reported to reduce the oxidation rates[9-12]. Conversely, other workers indicted an increase in the oxidation rate when the concentration of silicon was greater than 1.5wt%[9,13]. Silicon, apart from modifying the oxidation rates, is also effective in preventing titanium from stratifying[6,7]. But for silicon, being an α stabiliser, a great quantity of it may impair the mechanical performance of the resulting surface alloy.

In this paper, we describe the irradiation of the surface of Ti-6Al-4V samples precoated with aluminium, silicon or a mixture of aluminium plus silicon. The effects of these elemental additions on microstructure, microhardness, isothermal as well as cyclic oxidation are discussed.

2 EXPERIMENTAL PROCEDURE

Powders of silicon (d~160μm) and aluminium (d~20μm) were suspended separately or together in ethanol to form a slurry. This slurry was then sprayed on the polished titanium samples of 1.5 x 1.3 x 0.25 cm dimensions. The thickness of these predeposits varied between 26μm (Si only) to 40μm (Si+Al mixture). The samples were then placed on a computer-controlled x-y translation stage with a maximum scanning speed of 40 mm.s^{-1} and a resolution of 1μm. A continuous wave Nd:YAG laser (maximum power, 350 Watts) equipped with an optical fibre for guidance was used. The focalised spot diameter was 950μm. Irradiations were performed under an argon shield to prevent unwanted oxidation of samples during the treatment.

After varying the scanning speed and the incident power in order to determine the optimum condition of elaboration, the following operating parameters were adopted:

power   (P) = 150W
scanning speed (V_b) = 4 mm.s^{-1}
overlap   (δ) = ~50%
interaction time   (τ) = 0.238s
power density (q_0) = 2.1 x 10^8 W.m^{-2}

3 CHARACTERIZATION OF THE COATINGS

X-ray Diffraction Analysis

Following irradiation performed on samples containing aluminium or silicon separately, the analysis detected the presence of Ti_3Al and $TiAl_3$ or Ti_5Si_3 and α-Ti respectively. In the surface alloy containing both elements, all the above mentioned phases were detected except $TiAl_3$.

Composition of the Surface Alloys Formed

The results of the electron probe microanalyses performed on the alloys indicated a value of (i) 40 at%Al (ii) 12 at%Si and (iii) 20 at%Al + 9.5 at%Si for alloys containing only aluminium, silicon, or aluminium plus silicon respectively. Vanadium was always measured to be 4at%. From this point onwards, the alloys will be referred to as Ti-40Al-4V, Ti-10Al-12Si-4V and Ti-20Al-9.5Si-4V. The composition of the substrate in atomic percent can be represented as Ti-10Al-4V.

Surface Examination

Visual observation showed the surface to be shining with alternating ripples. Ripple formation is characteristic of surfaces treated with high energy beams[14-16]. This can be explained by the effect of surface tension gradient ($d\sigma/dT$) that existed on the surface during irradiation, an effect which is a consequence of the localized nature of the laser beam. The temperature at the centre of the beam being higher than at the edges, the surface tension increases from the centre of the molten pool to the edges. This creates a force that drags the liquid from the centre to the edges with the

consequence that the centre is depressed while the edges
are elevated.

In order to estimate the extent of the surface
undulations, confocal laser microscopy was used to measure
the ripple height, which is an indication of the surface
roughness. Figures 1(a) to 1(c) show the results of the
measurements where it can be seen that Ti-40Al-4V has the
maximum ripple height of the 85μm followed by the 26μm for
Ti-20Al-9.5Si-4V. When only silicon was added, i.e. with
Ti-10Al-12Si-4V, the surface was relatively flat after
irradiation. Considering the relative values of the
surface tension gradient of aluminium and silicon (-0.35
and -0.13 mN.m^{-1}.K^{-1} respectively)[17], this observation is not
surprising since the dragging force due to the effect of
the surface tension gradient will be higher for aluminium
than for silicon. When aluminium and silicon are present

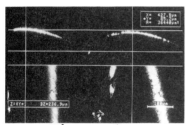

Figure 1a

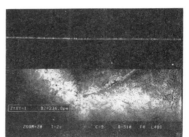

Figure 1b

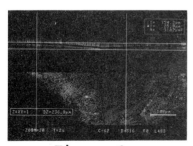

Figure 1c

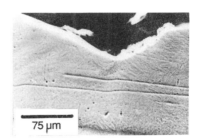

Figure 2

Figure 1 Results of the ripple height measurements with
          confocal laser microscope; (a) Ti-40Al-4V,
          (b) Ti-10Al-12Si-4V and (c) Ti-20Al-9.5Si-4V
Figure 2 SEM micrograph of the traverse section of Ti-
          40Al-4V; the shape of the ripple observed by
          laser microscope is confirmed

simultaneously in the surface alloy, a reduction in this
dragging force by silicon is observed compared to samples
without silicon. The result is a decrease in the ripple
height.

Microstructure

SEM observations of the transverse section of the
alloys showed the dependence of the microstructure on the
elements added. Figure 2 is a micrograph of Ti-40Al-4V
with fine needle-like crystals of $TiAl_3$ and Ti_3Al
interspersed in a surface alloy of 100µm thickness. On
the other hand, Ti-20Al-9.2Si-4V and Ti-10Al-12Si-4V
showed dendritic microstructures with Ti_5Si_3 found mainly
along the inter-dendritic arms. The melt depths were 200
and 160µm respectively, as shown in Figure 3. In all the
surface alloys, the solidification commenced with a narrow
plane front solidification at the alloy/metal interface,
followed by needle or dendritic microstructure.

In such solidification processes, the obtained
microstructure depends largely on the G/V [18] ratio (G =
temperature gradient, V = solidification rate). If this
ratio is greater than a certain critical value, plane
front solidification will result. Laser treatment is
characterized by the fact that at the interface where
solidification starts, the growth rate is almost
negligible while the temperature gradient is very high.
This gives rise to an infinite G/V ratio with
corresponding plane front and as the solidification
progresses, V increases and G/V decreases giving rise to
perturbations and dendritic microstructures as observe
experimentally.

Another peculiarity observed in Ti-10Al-12Si-4V is
the difference in microstructures between the central and
the overlapped region of the laser track. This is shown
in Figure 4. The concentrations of the two regions
measured by EDS were in good agreement with an eutectic
phase for the central region with α-Ti and Ti_5Si_3 finely
interspersed while the overlapped region consisted mainly
of hypo-eutectic structure with α-Ti grains and a mixture
of α-Ti and Ti_5Si_3 at the grain boundaries. This had been
attributed to the temperature difference existing between
the two regions during laser treatment[19].

Microhardness

Microhardness tests were performed on cross-sections of Ti-20Al-9.5Si-4V and Ti-10Al-12Si-4V. The results, presented in the Figure 5, showed that the hardness variation with depth at a constant load of 100gf for the two alloys was approximately the same except that Ti-20Al-9.5Si-4V was slightly harder and that the hardness of Ti-10Al-12Si-4V was relatively more uniform. This is probably due to the more uniform distribution of silicon in this last surface alloy.

Another useful important parameter relative to microhardness is the "Indentation Size Effect". This is simply a means of observing the dependence of hardness on the applied load. Loads varying from 50 to 500gf were applied on the surface and the indentation diagonal measured. The result, as depicted in Figure 6 showed the beneficial effect of elemental addition on hardness. In the Figure, it can also be seen that aluminium additions change the deformation processes of the surface alloys which become less hard when load increases. This indicates predominantly grain boundary deformation.

4 ISOTHERMAL OXIDATION TESTS

Kinetic Results. The samples were oxidized in static air at a pressure of 1 atm. The temperature of oxidation lay between 1173 and 1273 K.

Comparative results, shown in Figure 7 indicated a great improvement in the oxidation resistance for the three alloys compared to unalloyed Ti-6Al-4V, Ti-20Al-9.5Si-4V exhibiting the best behaviour.

When compared with bulk Ti_5Si_3[11], Ti-20Al-9.5Si-4V recorded a slightly higher weight gain. This can be explained by the presence of α-Ti (which has a high oxidation rate) in the elaborated surface alloys and also by the increase in surface area due to ripple formation. With a ripple height of about 26μm measured by confocal laser microscopy, the approximate increase in surface area is about 10%. Therefore, the results presented here should be taken as the most pessimistic as far as the kinetic curves are concerned.

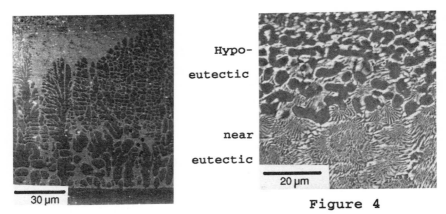

Figure 4

Figure 3

Figure 3   Typical SEM micrograph of the dendritic growth
            on Ti-20Al-9.5Si-4V and Ti-10Al-12Si-4V. Note
            the refinement of the microstructure near the
            surface.

Figure 4   The surface of Ti-10Al-12Si-4V showing the hypo-
            and near-eutectic structures

     For all the temperatures investigated, the kinetics
laws were parabolic. Activation energies of 260 and 195
kJ.mol^{-1} were calculated for Ti-20Al-9.5Si-4V and Ti-10Al-
12Si-4V respectively. These values are in agreement with
those reported by others[20] who worked on Ti-Si systems.

     Composition and Morphology of the Oxide Scales. TiO_2
(rutile) and Al_2O_3 (corundum) were observed by XRD in the
oxide scales irrespective of the temperature of oxidation.
Silicon was detected by EDS but silica was not detected by
XRD. This signifies probably its existence as an amorphous phase. As far as the oxide formation is concerned,
vanadium was not detected. This may be due to its low
concentration in the formed surface alloys.

     Surface Examination. Surface observations made by
SEM after oxidation showed that the degree of adherence
and compactness differed for the three alloys under
consideration. Ti-20Al-9.5Si-4V appeared to be more
adherent. Ti-40Al-4V showed a great tendency to spall
after cooling to room temperature. In some areas where
the oxide scale spalled on Ti-10Al-12Si-4V, observation
showed a case of preferential oxidation where the primary
α-Ti grains had been oxidized more than the grain boundaries where α-Ti and Ti_5Si_3 coexisted[19]. This region

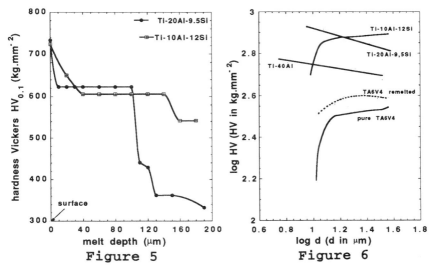

Figure 5

Figure 6

Figure 5 Comparison of hardness profiles
Figure 6 Comparison of the variation of hardness with
        applied load for different surface alloys

of preferential oxidation is estimated to be about 5% of
the total surface area. This phenomenon was not observed
in Ti-20Al-9.5Si-4V.

Transverse Section of the Oxide Scale. SEM
observations on Ti-10Al-12Si-4V and Ti-20Al-9.5Si-4V
showed the oxide scale to consist of three sublayers as
depicted in Figures 8(a) and 8(b). The three layers are:
(i) an external layer that consisted of a mixture of Al_2O_3
and TiO_2 in the case of Ti-10Al-12Si-4V, on the Ti-20Al-
9.5Si-4V, only Al_2O_3 was observed in this layer; (ii) the
intermediate layer that consisted of SiO_2 and TiO_2 in the
two alloys; (iii) an internal layer in contact with the
unoxidized alloy that consisted of a thin layer of pure
SiO_2 in Ti-10Al-12Si-4V and a high concentration of
unoxidized aluminium and silicon in Ti-20Al-9.5Si-4V.

The effect of temperature on the oxide scale was
shown to be different for the two surface alloys. While
the thickness of the external scale increased with
increasing temperature on Ti-10Al-12Si-4V, a reduction was
observed in the case of Ti-20Al-9.5Si-4V. This was
confirmed by the measurement of the external layer
thickness which is 12.5μm for the former and 6μm for the
latter, with the same oxidation temperature of 1173K.
This distinction could be explained by the presence of TiO_2

in the external scale of Ti–10Al–12Si–4V, which may be a consequence of insufficient aluminium content. TiO_2, acting as fast transport channels[7,8], would permit titanium cations to diffuse to the surface to form TiO_2, thereby increasing the external layer thickness. The high aluminium and silicon content of Ti–20Al–9.5Si–4V prevented the formation of TiO_2 in the external layer, keeping the thickness almost constant.

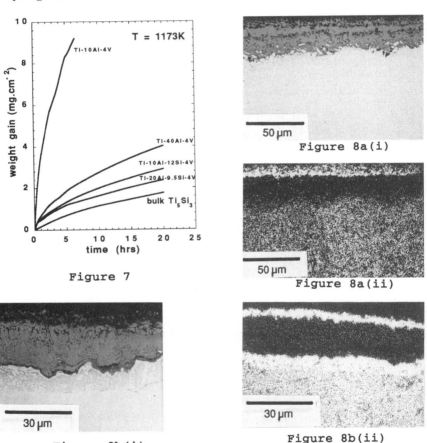

Figure 7

Figure 8a(i)

Figure 8a(ii)

Figure 8b(i)

Figure 8b(ii)

Figure 7   Comparison of isothermal oxidation tests between the surface alloys elaborated and bulk Ti_5Si_3

Figure 8   Transverse section of the oxide scale formed at 1183K during 100 hours exposure
(a) Ti–10Al–12Si–4V (i) Backscattered image (ii) Al map
(b) Ti–20Al–9.5Si–4V (i) Backscattered image (ii) Al map

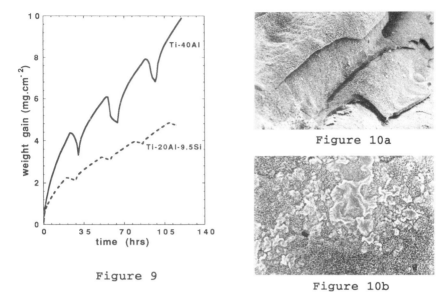

Figure 10a

Figure 9

Figure 10b

<u>Figure 9</u>   Comparison of cyclic oxidation tests between Ti-
            40Al-4V and Ti-20Al-9.5Si-4V
<u>Figure 10</u> Surface aspect after cyclic tests; (a) Ti-40Al-
            4V and (b) Ti-20Al-9.5Si-4V

Looking at the relative position of the oxide layers,
we can therefore say that the scale morphology is
compatible with the simultaneous inward diffusion of
oxygen and outward diffusion of Al^{3+} and Ti^{4+}. This
explains the parabolic kinetics observed. The presence of
SiO_2 promotes the formation of an external alumina layer
that blocks the inward diffusion of oxygen partially or
totally, depending on its degree of coherence and
compactness, which in turn depends on its concentration in
the surface alloy.

5 CYCLIC OXIDATION

Since the aim of the present study was to form surface
alloys that will not only be reproducible and chemically
homogeneous, but will also withstand aggressive
environments by forming a compact and adherent oxide
scale, we performed cyclic oxidation tests on Ti-20Al-
9.5Si-4V (the best surface alloy formed) to observe the
influence of the added elements on the adhesion of the
oxide scale. For the purpose of comparison, Ti-40Al-4V
was put under the same test, which was performed at a

plateau of 1183K. This plateau was held for 24 hours and four cycles were performed on each of the alloys. The heating and the cooling rates were 1173 and 723 K/hour respectively.

Figure 9 represents the results from the test. Ti-20Al-9.5Si-4V lost about 10 times less weight than Ti-40Al-4V, indicating much better cyclic oxidation behaviour. SEM studies of the surface after cyclic testing presented in Figures 10(a) and 10(b) confirmed these observations.

6 CONCLUSIONS

We have shown that laser irradiation of precoated Ti-6Al-4V led to the formation of crack-free and chemically homogeneous surface alloys. All the three alloys studied exhibited improved high temperature oxidation resistance.

Generally by modifying the surface-related properties, a compromise must be reached between the mechanical and chemical properties. In Ti-20Al-9.5Si-4V, we have gained on both sides since the oxidation resistance as well as the hardness and adherence were superior to those of the base alloy. In terms of surface finishing, aluminium increased the surface roughness with the worst adherence, while silicon eliminated it with average adherence. The combination of both aluminium and silicon increased the adherence, hardness, isothermal as well as cyclic oxidation resistance, with minimum rugosity. As far as the oxidation resistance was concerned, the presence of SiO_2 promotes the formation of an external alumina layer which was responsible for the superior behaviour of the surface alloy. It is further demonstrated that the addition of silicon (i) reduces the amount of aluminium necessary for the formation of a continuous Al_2O_3 layer on Ti-6Al-4V and (ii) allows the formation of SiO_2 which contributes to the plasticity of the scale as observed by cyclic oxidation test.

ACKNOWLEDGEMENT

This is part of the Ph.D work of one of the authors, A. Y. Fasasi. He is thankful to C.N.O.U.S. (Centre National des Oeuvres Universitaires et Scolaires) for the award of a four year scholarship.

REFERENCES

1.   T. Chande and J. Mazumder, Opt. Eng., 1983, 22, 362.
2.   S. Jansen and S. Gulbransen, 4th Int. Cong. Met.
     Corr., Amsterdam, 1969.
3.   R.A. Perkins, K.T. Chiang, G.H. Meier and R. Miller,
     'Oxidation of High Temperature Intermetallics',
     Edited by T. Grobstein and J. Doychak, TMS
     Warrendale, PA, 1989, p.157.
4.   H.A.F. ElHalfawy, E.S.K. Menon, M. Sundararaman and
     P. Mukhopadhyay, Proc. 4th Int. Conf. on Titanium,
     Edited by A. Kimura and O. Izumi, 1980, p. 1379.
5.   K.E. Weidman, S.N. Sankaran, R.K. Clark and T.A.
     Wallace, in reference 3, p. 195.
6.   H. A. Lipsitt, Mat. Res. Soc. Symp. Proc., 1985, 39,
     351.
7.   G. Welsch and A.I. Kahveci, reference 3, p.207.
8.   G.H. Meier, B. Appalonia, R.A. Perkins and K.T.
     Chiang, reference 3, p.185.
9.   C.J. Rosa, Oxid. Met., 1982, 17, 359.
10. H.W. Maynor and R.E. Swift, Corrosion, 1956, 12, 293.
11. A. Abba, A. Galerie and M. Caillet, Ann. Chim. Fr.,
     1979, 4, 15.
12. A. Gannouni, A. Galerie and M. Caillet, Ann. Chim.
     Fr., 1983, 8, 191.
13. G.P. Nadutenko, S.A. Gorbunov, I.S. Anitov and V.P.
     Teodorovich, Rev. Met. Lit., (ASM Metal Park Ohio),
     1967, 540, 2439.
14. P.R. Strutt, Mat. Sci. Eng., 1980, 44, 239.
15. H.E. Cline, J. Appl. Phys., 1981, 52, 443.
16. T.R. Anthony, H.E. Cline, J. Appl. Phys., 1977, 48,
     3888.
17. Smithells' Metals Reference Book, 6th Edition, Edited
     by E.A. Brandes, Butterworths, London, 1983, p.14-7.
18. W. Kurz and D.J. Fisher, 'Fundamentals of Solid-
     ifications', Trans Tech Publications, Switzerland,
     1986, Chapters 3 and 4, p.47.
19. A.Y. Fasasi, S.K. Roy, A. Galerie, M. Pons and M.
     Caillet, Mat. Lett., in print.
20. A.M. Chaze and C. Coddet, Oxid. Met., 1987, 27, 1.

Section 1.2 Aqueous Corrosion

1.2.1
Improved Powder Compositions for Thermochemical Diffusion Treatments of Metals and Alloys

A. D. Zervaki,[1] G. N. Haidemenopoulos,[1] and D. N. Tsipas[2]

[1] MIRTEC S.A., VOLOS, GREECE

[2] ARISTOTELES UNIVERSITY OF THESSALONIKI, PHYSICAL METALLURGY LABORATORY, DEPARTMENT OF MECHANICAL ENGINEERING, THESSALONIKI, GREECE

1   INTRODUCTION

This paper deals with the problem of high temperature oxidation and thermal fatigue properties of metals and alloys and the thermochemical treatments used to improve these properties.

It refers to an improved powder mixture composition, which allows introduction of rare earth elements simultaneously with other elements such as Al, Cr and B by thermochemical treatment into the surface of metals and alloys. Using this novel powder mixture composition, great improvements are obtained in the high temperature cyclic oxidation properties of treated metal and alloy substrates.

The current methods for protecting Ni-base superalloys operating at high temperature and oxidizing environments include thermochemical aluminizing, or chromizing by the pack cementation method. [1]

The pack cementation powders currently used contain a substance which releases the donor element X, where X = Cr or Al, to be diffused into the surface of the component under treatment, an activator or carrier which is usually in the form of a halogen containing compound, such as chloride or fluoride, and promotes the release of X, and an inert dilution compound, the dilutant, with refractory properties (e.g. Al_2O_3).

At present at least one component belonging to each of these classes of substances essential for the pack cementation thermochemical diffusion process, has to be added in the form of a powder mixture composition.

Utilizing this powder mixture composition the process is carried out in a furnace by heating the workpiece at an appropriate temperature and with the surface to be treated in contact with the pulverulent composition and holding the workpiece at this temperature for a suitable time.

During the process either Cr or Al diffuse into the surface of the alloy forming intermetallic compounds with improved high temperature oxidation/corrosion properties.

In this effort to improve the high temperature oxidation/corrosion properties, especially under thermal cycling conditions, the addition of small concentrations of certain "active elements" such as Yt and the rare earths, when added into the bulk of the alloy composition, have been found to be extremely beneficial. However these active elements when added in the bulk have a marked influence on workability, this latter being extremely sensitive to concentration. [2]

In order to counteract this negative effect on workability, it has been here attempted and achieved to introduce these active rare earth elements simultaneously with other elements such as Al and Cr during thermochemical treatment into the surface of a Ni-base alloy in an one-step process.

2   EXPERIMENTAL PROCEDURE

The pack cementation powder mixture used in our experiments contained : Hf / halide activators / Al_2O_3 / Al-supplier, or Hf / HfO_2 / halide activators / Al_2O_3 / Al-supplier, in different proportions for obtaining the Hf modified aluminide coatings and Al-supplier / Al_2O_3 / halide activators for the aluminide coatings.

The appropriate powder mixture was placed in a retort in contact with the INC 718 specimens and heated to a temperature above $800^{\circ}C$ for a period between 4-6 hours in an inert atmosphere, or under normal atmospheric conditions.

The obtained diffusion layers and cyclic oxidation products were characterized by optical and electron microscopy, EDAX and X-Ray diffraction.

Thermal cyclic oxidation tests were carried out in a specially constructed rig which allowed specimens to be cycled between high temperatures and room temperature every half hour. Each cycle consisted of heating the material at $1150^{\circ}C$ for 30 minutes and then cooling it at room temperature.

3   RESULTS AND DISCUSSION

It has been observed [3] that small amounts of rare earth oxides, or their pure metals or both, when added to the pack cementation powders used for aluminizing and chromizing, can provide an extremely effective powder composition for enriching the surface of the treated workpiece with more than one element simultaneously, one being a rare earth element and the other element being one of Al or Cr. In Table 1 the major phases present after each thermochemical treatment are indicated.

Table 1 - XRD Measurements of Al and Al/Hf Coated INC718

MATERIAL	MAJOR PHASES PRESENT
Al coated INC718	AlNi
Al/Hf coated INC718	AlNi, $Al_{16}Hf_6Ni_7$, Hf

Specifically it has been found that by adding about 3 wt% of HfO_2, or Hf, or both HfO_2 and Hf, to a standard aluminizing pack cementation powder containing an Al releasing substance, an activator, such as NH_4F, or NH_4Cl and an inert dilution compound, such as Al_2O_3, the surface of the treated workpiece, a Ni alloy, was enriched with both Al and Hf during thermochemical treatment.

With standard pack cementation powders usually the diffusion of Al or Cr can be obtained. Using this improved powder composition rare earth elements can also be diffused into the surface of treated workpieces and in combination with Al or Cr significantly improve the oxidation/corrosion properties of metals and alloys.

Similar results in the diffusion layers morphology, nature of obtained layers and the surface properties, such as oxidation, especially under thermal cycling conditions, cannot be obtained by the standard pack cementation powder for aluminizing and chromizing.

To demonstrate the described process and the improvements in thermal cycling properties, the microstructure of standard aluminizing pack cementation layer on a Ni alloy, INCONEL 718 is presented in Figure 1a, and a layer enriched with both Al and Hf during thermochemical treatment with the new proposed powder mixture on the same alloy is presented in Figure 1b.

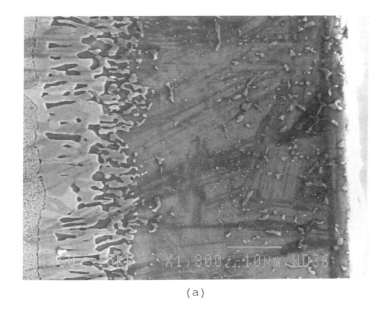

(a)

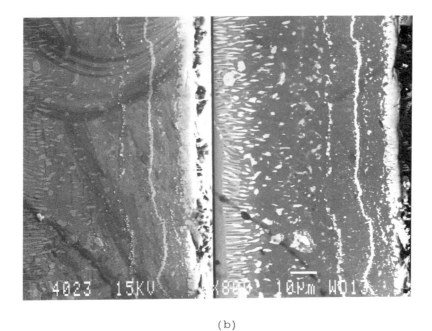

(b)

Figure 1 Typical microstructures of: (a) aluminized and (b) Hf modified aluminized coatings on a Ni-base superalloy

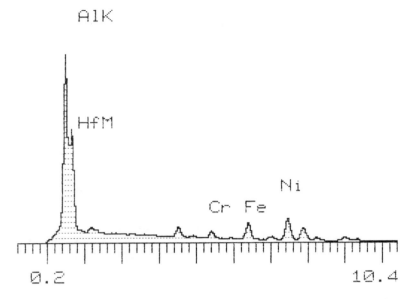

<u>Figure 2</u> EDX microanalysis of the white intermediate layer
in Hf-modified aluminized coatings showing the
presence of Hf

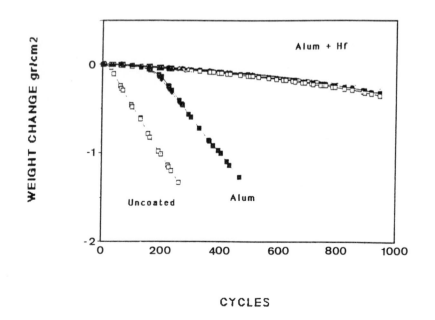

<u>Figure 3</u> Thermal cyclic oxidation behaviour of untreated,
aluminized and Hf modified aluminized coating on
INCONEL 718 alloy

The Hf rich layer(s) is clearly indicated in the same figure and confirmed by EDX, in Figure 2. In Figure 3, thermal cycling oxidation data for the obtained Hf modified layer in comparison with the untreated alloy and standard aluminized material, are presented.

Also in Table 2 the phase observed in the cyclic oxidation products after 222 cycles, for the aluminized and Hf-modified aluminized coating, are indicated.

The nature and morphology of oxidation products explain the differences observed in the cyclic oxidation properties between the two treatments. The presence of Hf in the diffusion layer increases the oxide layer adhesion and alters the nature and morphology of the products.

Table 2 Oxidation products XRD measurements after 222 cycles

MATERIAL	MAJOR PHASES PRESENT
Al coated INC718	Al_2O_3, NiO, Cr_3O_8, Cr_2O_3 $NiCr_2O_4$, $NiFe_2O_4$, NiCrFe (alloy)
Al/Hf coated INC718	Al_2O_3, HfO_2, AlNi

4   CONCLUSIONS

1. A new powder mixture composition was developed which allows introduction, simultaneously and in a one-step process, of rare earth elements and Al or Cr, by thermochemical means into the surface of metals and alloys of Ni.

2. The new powder mixture contains, besides the standard
    constituents of pack cementation powders (which are:
    the donor, the activator and the dilutant), a small
    amount of a rare earth oxide, rare earth metal or
    both.

3. Using this new powder mixture for carrying out
    thermochemical treatments, rare earth elements can be
    introduced into the surface of Ni alloys such as INC
    718 thus avoiding the negative effects in the
    workability of alloys when these elements are
    introduced into the bulk composition.

4. The Al diffusion layer enriched with rare earth
    elements shows excellent oxidation / corrosion
    properties, especially under thermal cycling
    conditions.

Acknowledgements

The authors are grateful to the EEC for financial
support of this work through the BRITE/EURAM project.

REFERENCES

1   R. Kossowsky, S.C. Singhal, Editors, "Surface
    Engineering, Surface Modification of Materials",
    Nato ASI Series, Martinus Nijhoff Publishers, 1984.

2   E. Lang, Editor, " The Role of Active Elements in
    the Oxidation Behaviour of High Temperature Metals
    and Alloys", Elsevier Applied Science, 1989.

3   R. Bianco, M. Harper, R.A. Rapp, Journal of Metals,
    Nov. 1991, p. 68.

1.2.2
The Electrodeposition and Corrosion Properties of Zinc–Chromium Alloys

Y. J. Su,[1] A. Watson,[2] M. R. El-Sharif,[1] and C. U. Chisholm[1]

[1] SCHOOL OF ENGINEERING, THE QUEEN'S UNIVERSITY GLASGOW, UK

[2] DEPARTMENT OF CHEMISTRY AND CHEMICAL ENGINEERING, PAISLEY COLLEGE, UK

1 INTRODUCTION

In recent years much interest has been shown in the deposition of zinc alloys with for example nickel and cobalt for their improved corrosion protection properties over single metal zinc deposits[1,2]. Zn-Ni alloys of 12% to 15% Ni and Zn-Co alloys with a small quantity of Co, 0.2% ~ 1.2%, are widely viewed as more resistant to corrosion in comparison to zinc coating[1,2]. These alloys however generally only achieve their superior performance after formation of conversion coatings[3]. This usually involves the use of highly toxic, carcinogenic and environmentally undesirable chromium(VI) solutions. The conversion coating formed contains complex chromium(III) and (VI) oxide species. Previous attempts to produce conversion coatings from less toxic materials, such as by the anodic and cathodic polarization of zinc in molybdates and tungstates solutions[4,5] have given passivated films that have shown a certain degree of corrosion protection, but generally less effective than chromium oxide-containing conversion layers. It was therefore decided to investigate the deposition and properties of zinc-chromium alloys, in which the chromium species of the conversion layer can be generated from within the deposit itself.

The electrodeposition process described in this paper yields high quality zinc alloy coatings containing 3-10% of chromium, with an optimum chromium content of 4-6%. Conversion coatings can be formed on these alloys by dipping in certain simple oxidants, free of chromium, which are environmentally more acceptable than the conventional chromium(VI) solutions. The corrosion resistance of these alloys, in the as-plated state,

without the conversion coating is somewhat better than
conventional acid zinc, although not markedly. The
conversion coatings produced on zinc-chromium alloys with
oxidants such as permanganate appear very similar to those
produced on acid zinc with conventional chromate solutions
and the corrosion resistance is very similar.

The advantage of the new process is a conversion
coating process free of highly toxic chromium(VI). The
electrodeposition process does involve chromium but in the
form of the environmentally much more acceptable chromium
(III). Repair of the conversion coating is possible with
chromium-free solutions.

The codeposition of the Zn-Cr alloy is based on an
acid zinc bath containing trivalent chromium(III). The
acid zinc plating process is among the most important
rapidly replacing the traditional cyanide process and has
been much modified to overcome its original poor throwing
power and insufficient brightness[7].

A 'trivalent' or chromium(III) bath offers an
environmentally more acceptable process but until recently
the thickness of the coatings obtained remained only a few
microns and sustained deposition was only recently
achieved[8]. It is now possible to produce thick chromium
and chromium alloy coatings from chromium(III) amide
electrolytes or a number of chromium(III) aqueous
electrolytes such as urea and carboxylic or amino-acid
electrolytes[9]. The present work has extended these
techniques to the codeposition of Zn-Cr alloys.

2 EXPERIMENTAL

Analytical grade chemicals were used in the electrodeposition and the corrosion tests. For the electrodeposition mild steel cathodes with a dimension of
2.5x2.5cm were polished with silicon carbide paper of up
to grade 500. They were then treated cathodicly in a
solution containing 50~70g/l alkaline cleaning agent with
a current of 5 A/dm^2 for one minute at 80^{0}C, followed by a
rinse in deionised water before plating. High density
graphite anodes were used in the electrodeposition of Zn-
Cr alloys. In the studies of corrosion resistance
properties the types of coating examined were a commercial
acid Zn, a commercial chromated acid Zn, a laboratory

prepared Zn-Cr, and a laboratory prepared Zn-Cr with a conversion coating formed with permanganate or persulphate. These coatings were all selected to have a thickness about 8~10μm.

Zr-Cr alloy coatings were electrodeposited from a chromium(III) urea electrolyte containing secondary complexants and its optimal bath is shown in Table 1. In the formation of the conversion coatings the electro-deposited coatings were pretreated in a 3% nitric acid solution for 10 seconds and then rinsed in deionized water. This was followed by dipping the coatings into a solution containing permanganate, pH = 1.5~1.8, or persulphate, pH = 1.5~1.8 and at 25^0C. After dipping the coatings were rinsed again in deionised water and dried in 60^0C hot air flow for two hours.

Table 1 The optimal bath composition for the electro-
        deposition of Zn-Cr alloys

Component	Amount (g/L)
$ZnSO_4 \cdot 7H_2O$	57
$CrCl_3 \cdot 6H_2O$	215
Urea	240
*secondary complexant	38
H_3BO_3	9
NH_4Cl	27
NaCl	29
pH	2.5~3.0
Temperature,(°C)	20~25
Agitation	Cathode movement vertical

Conversion coatings obtained with permanganate have an appearance similar to a conventional heavy chromate conversion coating. Those with persulphate are similar in appearance to lighter conventional conversion coatings.

The composition of the alloys was determined by X-ray fluorescence in a JXA-50A Electron-microscope with a Link electron microprobe analysis unit. Coating thicknesses were determined with a Buechler Omnimet-1 Image Analysis System.

Electrochemical Investigation of the Deposits

In the studies of electrochemical properties a three electrode cell system was used in which the sample was used as the working electrode, with platinum and saturated calomel electrodes as the auxiliary and reference electrodes respectively. Each sample was degreased with acetone and coated with lacquer to give an exposed area of $4cm^2$ before immersion into a non-deaerated 1.0M NaCl solution of pH 6.0 at 30^0C. The values of the free corrosion potential of the samples were recorded with respect to a saturated calomel reference electrode for a period of 170 hours. Tafel plots were characterized with a potentiodynamic scan rate at 5mV/min., and the corrosion current for each coating was estimated by Tafel line extrapolation. The corrosion rates denoted by corrosion currents against time were obtained by the polarization resistance method. The polarization resistance, R_p, was determined from the slope at the origin of the polar-ization curve, as the Stern-Geary equation[10], relating the polarization resistance to the corrosion current

$$I_{corr} = (\beta_a \times \beta_c) \; / \; [2.303 \times R_p \times (\beta_a + \beta_c)] \qquad (1)$$

where β_a and β_c are the anodic and cathodic Tafel constants respectively and were determined by Tafel extrapolation. In the present case where dissolved oxygen exists in the solution it was assumed that the cathodic reaction is diffusion controlled. Thus the corrosion currents could be calculated from a simple form of equation (1) as

$$I_{corr} = \beta_a / \; (2.303 \times R_p) \qquad (2)$$

In the laboratory accelerated salt spray corrosion tests, 5% natural salt spray was used according to ASTM

Specification B117, with a Weiss Technik SSC450 salt spray
testing chamber.

3 RESULTS AND DISCUSSION

<u>Electrodeposition</u>

Chromium percentages in the zinc-chromium alloys are
shown in Figure 1 as a function of cathode potential with
different metal concentrations in electrolyte. It can be
seen that the chromium content in the alloys increases
with decreasing zinc content in the electrolytes, but
experiments showed that a low zinc concentration (0.1 mol/
l or below) would produce only thin deposits with black
streaks. To obtain thick alloy deposits the zinc level in
the electrolytes should be above 0.2 mol/l, although the
amount of chromium in alloys was generally found unacceptably low when the zinc level in the electrolytes is
higher than 0.5 mol/l. The optimal chromium content in
the electrolyte was found to be 0.8 mol/l, while contents
below 0.6 mol/l lead to a low chromium percentage in
deposits with poor quality.

At pH values below pH 2.0 severe hydrogen evolution
and low current efficiency were observed. The deposits
then showed pitted surfaces as a result of gas bubbles
sticking during deposition, despite the use of wetting
agents. This indicates that with a low ratio of
$[Zn^{++}]/[H^{+}]$ in the electrolyte the detachment of bubbles at
their initial stage of growth is a controlling factor in
Zn-Cr electrodeposition. Cathode movement vertically
improves the quality and is believed to favour the
detachment of hydrogen gas bubbles at the initial stage of
growth, due to the vertical movement of the detached
bubble stream.

A further increase of pH above 2.0 showed no
improvement in terms of deposit quality, but black streaks
start to appear as the bulk pH was increased beyond 3.5
due to local precipitation of hydroxochromium(III)
species. An increase in temperature generally resulted in
a decrease in current efficiency and chromium content in
the deposits. It also increases the rate of μ-
hydroxochromium(III) polymer formation, leading to
eventual failure of sustained deposition. Higher
temperatures are therefore believed to enhance the

undesirable influence of hydrogen evolution so that a
suitable range of temperature to obtain better quality
deposits is 20~25°C.

Table 2 Effects of pH and temperature on deposit
        composition and current efficiency, with the
        electrolyte composition shown in Table 1, and at
        cathode potential of -1.7V(SCE)

pH	Temperature (^{0}C)	Chromium (%)	Current Efficiency (%)
1.5	25	3.3	31.43
1.8	25	3.8	37.93
2.0	25	4.2	42.58
2.5	25	4.5	48.13
3.0	25	4.0	47.14
3.5	25	3.0	37.71
2.5	20	4.2	44.02
2.5	30	3.6	40.86
2.5	35	3.2	32.35
2.5	40	3.0	30.56

Formation of the Conversion Coating

    As Zn-Cr alloy coatings are immersed into a
relatively acidic permanganate or persulphate solution
(pH=1.5~1.8) zinc is preferentially dissolved by reaction
with H$^+$ ions, thus leading also to the dissolution of
chromium. Within the region next to the metal surface the
relevant reactions, with acid permanganate, might be
represented in a simplistic manner by:-

$$Zn + 2H^+ \rightarrow Zn^{2+} + H_2$$

$$5Cr + 3MnO_4^- + 24H^+ \rightarrow 5Cr^{3+} + 3Mn^{2+} + 12H_2O$$

$$10Cr^{3+} + 6MnO_4^- + 11H_2O \rightarrow 5Cr_2O_7^{2-} + 6Mn^{2+} + 22H^+$$

    These reactions result in the consumption of H$^+$ ions
and thus the pH in the metal-solution interface rises
rapidly. Gelation occurs and a complex chrome gel
containing zinc, chromium(III) and chromium(VI)
oxo/hydroxo species is precipitated on the metal surface

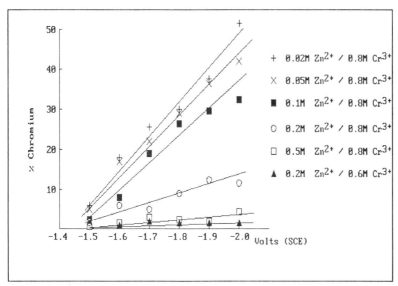

<u>Figure 1</u> The chromium percentage in electrodeposited zinc/chromium alloys against the applied cathode potential with varied metal ion concentrations in the electrolyte

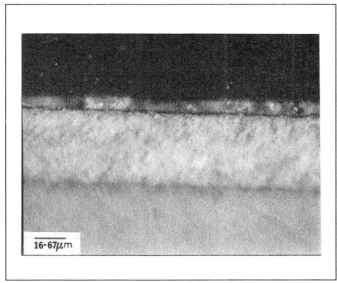

<u>Figure 2</u> A photomicrograph of a typical permanganated conversion coating

when the pH reaches a critical value. The appearance of
the conversion coatings obtained varied from light yellow
to iridescent depending on the operating conditions.
Coatings obtained with persulphate are generally lighter
in appearance than those with permanganate and somewhat
thinner.

Colour micrographs of the conversion coating surface
showed that the yellow-orange distribution in the film was
generally proportional to the chromium content in the Zn-
Cr alloy coatings. Under adequately controlled conditions
in the conversion process, thick layers of conversion
coating without microcracks can be obtained. Microcracks
are believed to reduce the protective value of the overall
coatings. Figure 2 shows the cross-section of a
permanganated conversion layer with a thickness of 6.25μm
on the alloy surface. The distribution of soluble and
insoluble portions of chromium oxides in the film is
thought to have an important influence on the corrosion
properties of the coatings. The investigation of that and
the relevant formation conditions will be reported
elsewhere.

Although the chemistry of the conversion coating has
not been investigated in any great detail, its properties
would seem as if the product is very similar to that from
the conventional process and involves soluble chromate
generated at the alloy surface.

Studies of the Corrosion Properties

Although conversion coatings on zinc-chromium alloys
can be obtained with permanganate or persulphate, those
from permanganate at this stage of development appear to
be better. Thus the results described in this paper will
concentrate on permanganated conversion coatings. The
results obtained with persulphate are broadly similar,
although the stability is at this stage somewhat poorer.

The change of corrosion potentials with time of
immersion is shown in Figure 3. It indicates that both
for Zn and Zn-Cr, without a conversion coating, the
potentials started at more active values and generally
shifted towards more positive direction. In contrast,
permanganated Zn-Cr and chromated Zn began with less
active values and their surfaces took a period of time to

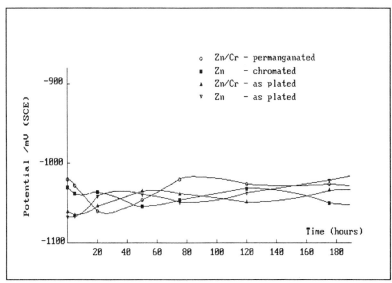

Figure 3 The potential against time of immersion in 1.0 M
NaCl solution at pH=6 and 30^{0}C; the zinc-chromium
alloy contains 6% chromium

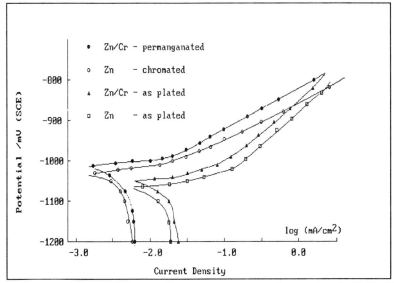

Figure 4 Polarization curves for coatings in 1 M NaCl
solution at pH=6 and 30^{0}C; the zinc-chromium alloy
contains 6% Cr

become activated. The potentials of chromated Zn become
less electronegative after about 20 hours, while the
permanganated Zn-Cr alloy did not change markedly in its
corrosion potentials. The range of potential variation
for both passivated (conversion coated) and unpassivated
Zn-Cr alloys suggest that they are both able to act as
sacrificially protective coatings for steel substrates
since their corrosion potentials were found to display a
large difference apart from the value for iron, that is,
-0.69V (SCE) under similar conditions. The permanganated
Zn-Cr conversion coatings have a somewhat greater
stability.

Figure 4 shows the anodic and cathodic polarisation
curves obtained after a one hour immersion test. The
estimation of the corrosion currents was made by Tafel
line extrapolation from these curves. It shows a lower
corrosion current for the Zn-Cr alloy than for single
metal Zn. The curves of permanganated Zn-Cr and chromated
Zn have a similar shape although a slightly, but
consistently, lower anodic dissolution for permanganated
Zn-Cr is shown. A comparison between these two coatings
suggests some improved corrosion resistance was offered by
the Zn-Cr alloy with permanganated conversion coating.
However the protection offered is broadly similar.

Prolonged immersion tests over 170 hours gave a
further comparison between the coatings. The results are
shown in Figure 5. It was noted that the corrosion rate
of single metal zinc was high during the first two days
and slowed down thereafter. This may be attributed to the
barrier effect on oxygen diffusion of adherent corrosion
products, effectively the basic hydroxide chloride[12] which
starts to develop and dominate the cathodic reaction with
time. This phenomenon of initial higher corrosion rate
was not observed for the Zn-Cr alloy, thus it may be
expected that the spontaneously formed layer on the alloy
surface has its own impact on the cathodic reaction. The
spontaneously formed oxide layer on the Zn-Cr alloy might
thus appear to be denser or more coherent than that on
zinc, and to offer a limited element of self passivation.
The spontaneously formed barrier on Zn-Cr will contain
zinc and chromium(III) species but no chromium(VI) and
thus be less effective than a conventional conversion
coating.

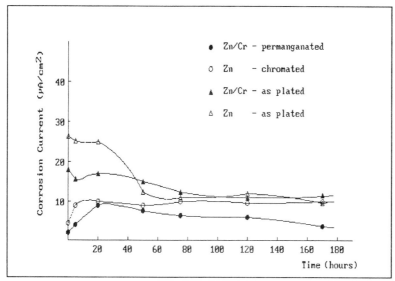

<u>Figure 5</u> The variation of corrosion currents against time
of immersion in 1.0 M NaCl at pH=6 and 30^{0}C for
zinc and 6% chromium-zinc

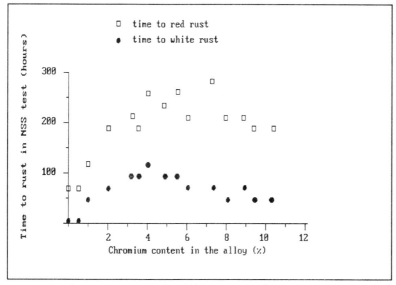

<u>Figure 6</u> Time to rust in NSS tests for permanganated zinc-
chromium alloys with various chromium contents

Permanganated Zn-Cr and chromated Zn show a similar tendency to each other in the prolonged immersion test, as can be seen in the curves in Figure 5, in which the corrosion rates initially reached higher values after a short period of time and then generally became lower. The similarity of their behaviour may be associated with a similar character of these coating surfaces. The initial increase in the corrosion rate is likely to be associated with the initial leaching out of soluble chromium(VI) species until equilibrium is reached. Again the permanganated Zn-Cr displayed slightly lower corrosion currents and greater stability, although the protection is broadly similar.

The results of 5% neutral salt spray (NSS) are shown in Figure 6. It gives the time to white and red rust for permanganated Zn-Cr alloy at various concentrations of chromium in the alloy. Deposits with chromium concentration between 4-8% possess significantly higher corrosion resistance. Below 4% there is insufficient chromium to form an adequate conversion coating on treatment with permanganate. Above 8% the metallic quality of the alloy deteriorates and this is also reflected in the permanganated coating. The optimum chromium content is between 4-6%.

The electrolyte shown in Table 1 has been further modified to introduce nickel to produce a ternary Cr-Ni-Zn alloy. This electrodeposited alloy can also form a high quality conversion coating by dipping in acidic permanganate. It has already been shown that in the as plated state the ternary alloy offers a much better corrosion resistance performance than the conventional acid zinc, or zinc-chromium alloys. This is in line with the improved performance of zinc-nickel alloys. Preliminary work would seem to indicate an enhanced performance following permanganating. Details of this work will form the basis of a future publication.

4 CONCLUSION

It is possible to produce zinc-chromium deposits with industrially applicable thicknesses from an acid zinc-based chromium(III) aqueous electrolyte. This process involves the use of certain techniques to ensure that sustained electrodeposition can be maintained. A low ratio of $[Zn^{++}]/[H^+]$ favours the codeposition of alloy but tends to encourage the simultaneous hydrogen evolution at the cathode. Excessive hydrogen evolution is detrimental to metallic quality. It was found that cathode movement vertically would be effective to reduce the influence of hydrogen gas bubbles. Thick deposits of high quality can be obtained with a chromium content up to about 10%. The introduction of chromium and zinc deposits shifts the corrosion potential somewhat towards more positive values, but the Zn-Cr alloy is able to act as sacrificial coating for steel and an improvement on corrosion protection over single metal zinc is evident. More significantly it provides an alternative way of generating a chromium oxide conversion film by a simple chemical oxidation in a chromate-free solution. Characteristically similar conversion layers to chromate can be formed by the immersion of Zn-Cr deposits into an acidic $KMnO_4$ solution. Electrochemical and accelerated corrosion tests indicate that permanganated Zn-Cr alloys, especially for those with chromium concentration between 4~6%, have a lower corrosion rate than commercial chromated zinc of the same thickness, but of a similar order. The advantage of the new process lies in the elimination of toxic chromate(VI) and its replacement by an environmentally more acceptable chromium(III) in the electrodeposition electrolyte.

REFERENCES

1. M.R. Lambert, R.G. Hart and H.E. Townsend, SAE Technical paper 831817, Soc. of Automotive Engineers, Warrendale, PA, 1984.
2. A.P. Shears, Trans. Inst. Metal Finishing, 1989, 67, 67.
3. N.R. Short, A. Abibsi and J.K. Dennis, Trans. Inst. Metal Finishing, 1989. 67, 73.
4. W. Paatsch, Metal Finishing, 1985, 79.
5. G.D. Wilcox and D.R. Gabe, Br. Corrosion J., 1987, 22, 254.
6. F.A. Lowenheim, "Electroplating - Fundamentals of Surface Finishing," 1987, 442-446.
7. T.V. Venkatesha, J. Balachandra, S.M. Mayanna and R.P. Dambal, Plating and Surf. Finishing, June 1987, 77
8. C.U. Chisholm, M.R. El-Sharif and A. Watson, 'Surface Engineering Practice,' 1990, 619.
9. A. Watson, A.M.H. Anderson, M.R. El-Sharif and C.U. Chisholm, Proceedings of Eurocorr., 1991, 1, 228.

10.   M.Stern, <u>Corrosion J.</u>, 1958, <u>13</u>, 440.

11.   A.M.H. Anderson, A. Watson, M.R. El-Sharif and C.U. Chisholm, <u>Trans. Inst. Metal Finishing</u>, 1991, <u>69</u>, 26-32.

12.   I.Suzuki, <u>Corrosion Sci.</u>, 1985, <u>25</u>, 1029.

1.2.3
Corrosion Resistance of Plasma Nitrided Cr-18Ni-9Ti Steel and Titanium

J. Mańkowski and J. Flis

INSTITUTE OF PHYSICAL CHEMISTRY, POLISH ACADEMY OF SCIENCES,
01-224 WARSAW, POLAND

1 INTRODUCTION

Nitriding is applied to a large variety of manufactured products to increase their wear and corrosion resistance, fatigue strength and surface hardness improving the tribological properties of steels.

The application of nitriding has been considerably increased since the development of the plasma technique, which differs from other nitriding techniques by using the electrical glow discharge to activate the nitrogen gas molecules needed for the process.[1] Plasma technique facilitates the removal of oxide films from the treated surface and hence it is especially advantageous for nitriding stainless steels and titanium.[2] Plasma nitriding produces a strongly bonded nitrogen compound zone and a diffusion zone actually integral with the original metal surface.

It has been reported [3-5] that corrosion resistance of stainless steel may be to an important degree affected by the nitriding process. Our previous measurements [6] have indicated a substantial decrease in corrosion resistance of stainless steels nitrided for 16 hours in mixture with 25% N_2; this decrease occurred especially within the diffusion zone. The change in the corrosion behaviour can be interpreted in terms of a high chromium depletion in the steel matrix due to the formation of chromium nitrides.[7-9]

By varying the process variables such as composition and pressure of the gas mixture, treatment time and temperature, it is possible to control the properties of nitrided layers.[10-11] By choosing appropriate nitriding conditions and metallurgical treatment it is possible to minimize the decrease in corrosion resistance or even increase this resistance.

Zhang and Bell [5] found that the decrease in corrosion resistance of nitrided stainless steels can be diminished by employing atmospheres with higher nitrogen contents and by carrying out the process at lower temperatures.

The objective of the present work was to determine the effect of nitriding on the corrosion resistance of austenitic stainless steel and titanium.

2 EXPERIMENTAL

Materials used were Cr18Ni9Ti austenitic steel (C 0.05, Cr 18.20, Ni 8.62, Mn 1.44, Si 0.22, S 0.018, P 0.039, and Ti 0.26 wt.%) and titanium (Al 0.045, Cr 0.025, Si 0.008, Fe 0.10, and C 0.015 wt.%). Specimens were in form of 3 mm thick discs. Before nitriding, Cr18Ni9Ti steel was solution annealed at 1050 °C.

Nitriding Procedures

Cr18Ni9Ti Steel. The steel samples were subjected to plasma nitriding at 585 °C for 6 hours or at 440 °C for 16 hours in a gas mixture containing 80 vol.% N_2 + 20 vol.% H_2 under a pressure of 670 Pa.

Titanium. Titanium was plasma nitrided at 860 °C for 9 hours in pure nitrogen under a pressure of 670 Pa.

Thickness and Structure of Nitrided Layers

Nitrided Steel. Microhardness profiles (Figure 1a) indicate that the nitrided layer growing at 585 °C was markedly harder and grew faster than that growing at 440 °C. From microhardness profiles it follows that the total thickness of nitrided layers obtained at 585 °C and 440 °C was about 70 and 48 μm, respectively. Plasma nitriding at 585 °C produced a thin outer zone composed of the ε + γ' nitrides (4 μm thick); a diffusion zone contained the γ' precipitates in deeper zones.

Nitrided Titanium. The total thickness of a nitrided layer, as estimated from the microhardness profile (Figure 1b), was about 30 μm. The thickness of the titanium nitride zone was very small (about 1 μm according to Roliński [12]). Changes in microhardness of the nitrided layer in titanium indicate that the content of nitrogen dissolved in the titanium matrix continuously decreased with the depth.

Corrosion Measurements

The corrosion resistance of nitrided materials was determined by potentiodynamic polarisation method in chosen solutions on as-nitrided and partially abraded surfaces. Anodic polarisation curves were measured by sweeping potential in the noble direction at a rate of 1 mV s^{-1}, starting from a potential of about 50 mV more negative than the corrosion potential. Potential was measured against a sulphate reference electrode, but is expressed here relative to the saturated calomel electrode (SCE).

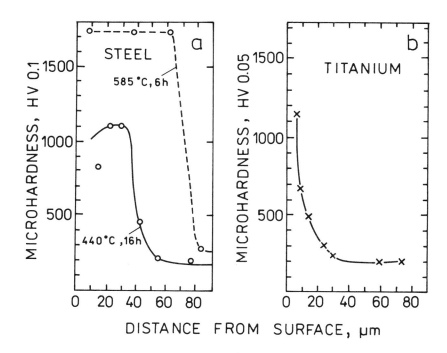

Figure 1 Microhardness profiles in Cr18Ni9Ti steel (nitrided in 80% N$_2$ + 20% H$_2$ gas at 585 or 440^{0}C) and titanium (nitrided in pure nitrogen at 860^{0}C)

The anodic behaviour of nitrided steel was examined at 25 °C in non-deaerated 0.05 M Na$_2$SO$_4$ acidified to a pH of 3.0, without chlorides and with 0.1 mole/l NaCl. Nitrided titanium was tested in 40% H$_2$SO$_4$ (at 40 °C).

3 RESULTS AND DISCUSSION

Anodic Behaviour of Cr18Ni9Ti Steel Nitrided at 585 °C

Anodic Behaviour in 0.05 M Na$_2$SO$_4$ (pH 3.0). In the solution of 0.05 M Na$_2$SO$_4$ the unnitrided steel is passive in a wide range of potentials from −0.4 to + 0.7 V (Figure 2). The increase in current density at potentials above about 0.7 V is typical of the transpassive behaviour of stainless steels.

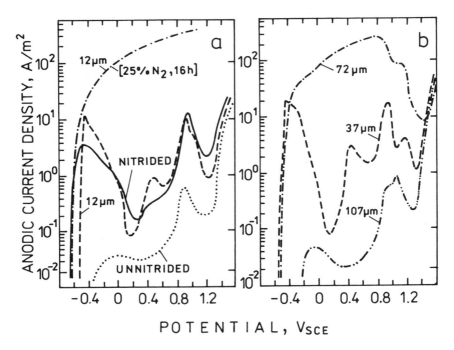

<u>Figure 2</u> Anodic polarisation curves for Cr18Ni9Ti steel
nitrided at 585⁰C for 6 h and abraded to given
depths. Test solution: 0.05 M Na$_2$SO$_4$ pH 3.0

For the nitrided steel the corrosion potential lay
in the active region and the active dissolution appears
in the wide range of potentials (Figure 2a). The polarisation curve for the surface nitrided in the mixture with
80% N$_2$ for 6 hours was similar to that for the steel nitrided in 25% N$_2$ for 16 hours which means that the ε nitride formed during the high-nitrogen nitriding possesses
better protective properties than γ' nitride formed during the low-nitrogen nitriding.

Anodic polarisation curves obtained for the partly
abraded surfaces of nitrided steel are shown in
Figures 2a and 2b. In the active region, the currents
were observed to slightly increase with the abrasion
depth. In the passive region, the currents at the depths
of 12 and 37 μm were comparable with those for the nitrided surface, showing the ability of this part of the diffusion zone to passivate. For comparison, the anodic
curve for the steel nitrided in 25% N$_2$ for 16 hours and
abraded to the same depth of 12 μm is plotted in
Figure 2a. It is apparent that after nitriding in 80% N$_2$
anodic currents are lower than those for the nitriding in
low-nitrogen atmosphere.

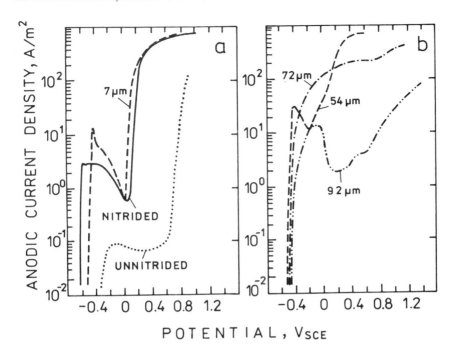

Figure 3 Anodic polarisation curves for Cr18Ni9Ti steel
          nitrided at 585°C for 6 h and abraded to given
          depths. Test solution: 0.05 M Na_2SO_4 + 0.1 M
          NaCl (pH 3.0)

The results of the present work, showing a
relatively small loss in corrosion resistance of a highly
nitrided steel, are in agreement with the earlier observations of Zhang and Bell [5] and Yan and Wang.[13] The former authors explained this fairly good resistance of
plasma nitrided 316 stainless steel by a preferential
formation of the γ' nitride which retards the precipitation of chromium nitrides in the steel nitrided. The
other possibility is that nitrogen dissolved in the metal
matrix may affect corrosion behaviour of nitrided material in a similar way as does nitrogen contained in
austenitic stainless steels.[14,15]

This relatively good corrosion resistance in
sulphate solutions occurs down to the depth of about
45 μm. However, anodic current increased with the increasing depth (Figure 2b). It means that a beneficial effect
of the nitrogen dissolved in a steel matrix does not
occur at these depths. However, the extent of chromium
nitride precipitation appears to be sufficiently high because the loss in corrosion resistance is considerable.

The extent of corrosion could probably be substantially decreased by the use of a high cooling rate after the nitriding process. Rayaprolu and Hendry [8] found that in the case of the cooling rate being sufficiently rapid to prevent precipitation, corrosion resistance of nitrided austenitic steel was not impaired.

It should be noted that after a complete removal of diffusion layers, the corrosion resistance of the treated steel is recovered (Figure 2b).

Anodic Behaviour in 0.05 M Na_4SO_4 + 0.1 M NaCl. In the solutions with chlorides, corrosion resistance of the steel studied was worse than that in the chloride-free sulphate solution of the same pH. Typically, pits developed during anodic polarisation. Onset of pitting was identified by a sharp increase in current density over a narrow range of potentials. E_{pit} was 0.65 V for unnitrided steel and 0.1 V for nitrided one. It can be seen from Figures 3 a and 3 b that an unnitrided surface shows higher resistance to chloride attack than nitrided and abraded ones. For abraded surfaces, the pitting potential decreased with the increasing abrasion depth within the nitrided layers. For the depth of 72 μm, an incessant increase in current density was observed, so that it was not possible to distinguish any pitting potential.

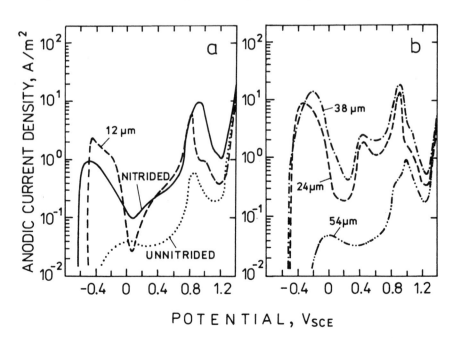

Figure 4 Anodic polarisation curves for Cr18Ni9Ti steel nitrided at 440⁰C for 16 h and abraded to given depths. Test solution: 0.05 M Na_2SO_4 (pH 3.0)

Anodic Behaviour of Cr18Ni9Ti Steel Nitrided at 440 °C

Anodic Behaviour in 0.05 M Na$_2$SO$_4$ (pH 3.0). Figure 4 presents polarisation curves measured in 0.05 M Na$_2$SO$_4$ solution on Cr18Ni9Ti steel nitrided for 16 hours at 440 °C. The anodic behaviour in sulphate solution is essentially the same as that shown in Figure 2 for the steel nitrided at 585 °C. The difference consists in a lack of a distinct increase in current densities for the deepest region of diffusion zone as it was observed in the case of steel nitrided at 585 °C.

Anodic Behaviour in 0.05 M Na$_2$SO$_4$ + 0.1 M NaCl (pH 3.0). An addition of chloride anions did not influence the polarisation behaviour in a significant way. Polarisation curves for nitrided steel (Figure 5) did not indicate any pit formation, except for the depth of 51 μm at which the nitrided layer was removed completely.

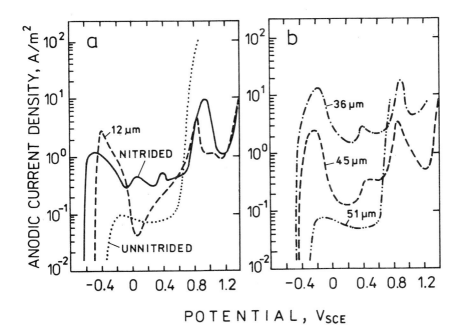

Figure 5 Anodic polarisation curves for Cr18Ni9Ti steel nitrided at 440°C for 16 h and abraded to given depths. Test solution: 0.05 M Na$_2$SO$_4$ + 0.1 M NaCl (pH 3.0)

The results of this work confirm the beneficial effect of the decrease in nitriding temperature on corrosion resistance, reported by other workers.[4,5,16-18] Zhang and Bell[5] found that after plasma nitriding of AISI 316

stainless steel at 400 °C there was no significant decrease in pitting potential relative to unnitrided steel.
Our measurements indicate that plasma nitriding at 440 °C
is even more advantageous in respect of the corrosion
resistance. This corrosion resistance of steel nitrided
at lower temperatures is ascribed by some researchers to
the formation of an undefined "S-phase".[18]   Mannehimer
and Paxton [4] suggested in 1967 that a formula $(Fe_3Ni)N$
could be assigned to the nitride phase formed at 400 °C.

Anodic Behaviour of Nitrided Titanium in H_2SO_4

Anodic polarisation curves shown in Figure 6
indicate that titanium exposed to a concentrated H_2SO_4
solution exhibits an active dissolution in the interval
of -0.85 to -0.50 V and an active to passive transition.

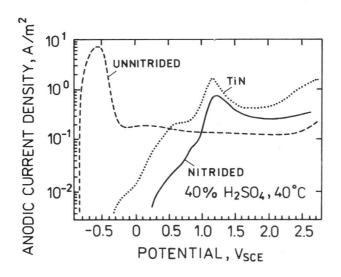

Figure 6 Anodic polarisation curves for nitrided titanium
        and sintered TiN

In contrast, the nitrided titanium yielded a
significantly more noble value of corrosion potential,
generally in the range of 0.2 to 0.4 V and did not exhibit an active to passive transition, thus indicating a
substantial increase in the corrosion resistance. The
beneficial effect of nitriding on the corrosion resistance of titanium can be explained by the formation of
a titanium nitride zone. The polarisation curve for the
nitrided titanium was similar to that for the sintered
TiN, demonstrating that corrosion behaviour of the nitrided surfaces is determined by the presence of TiN.

TiN is known to possess high hardness, low coefficient of friction, and high corrosion resistance in many corrosive environments. Although TiN is a fairly inert substance in its pure form, commercially applied TiN coatings may vary widely in their corrosion characteristics.[19,20]

After a slight abrasion of the nitrided titanium to the depth of about 8 μm (Figure 7), the nitride zone was removed and the diffusion zone was exposed. At this depth, the potential was lower than for the as-nitrided titanium (Figure 6). For further depths, corrosion potential shifted in the active direction and the anodic currents in the active region increased and tended to the value measured for the unnitrided titanium.

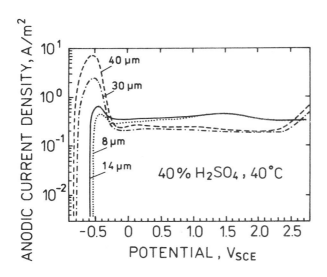

Figure 7 Anodic polarisation curves for titanium nitrided and abraded to given depths

It follows, therefore, that an increase in the corrosion resistance of the nitrided titanium is due chiefly to the formation of a titanium nitride surface zone. The corrosion resistance of the diffusion zone decreased with increasing depth, in parallel to the drop in microhardness and accordingly to the decreasing nitrogen content.

4 CONCLUSIONS

1. Cr18Ni9Ti steel nitrided in the mixture of 80% N_2 + 20% H_2 at 585 °C showed in 0.05 M Na_2SO_4 (pH 3.0) relatively good corrosion resistance for the outer ni-

tride zone and in the intermediate region of the diffusion
zone down to the depth of 45 μm. The deepest regions of
the diffusion zone underwent an intense anodic dissolution.

2. In the solution with chlorides, unnitrided Cr18Ni9Ti
   steel and also steel nitrided at 585 °C underwent pitting corrosion.

3. Low-temperature plasma nitriding (at 440 °C) resulted
   in a significant increase in the corrosion resistance
   of nitrided steel in comparison to nitriding at
   585 °C. In the solution with chlorides (0.1 M NaCl)
   corrosion pits did not develop.

4. Plasma nitriding of titanium in nitrogen resulted in
   an increase in the corrosion resistance in 40% H_2SO_4
   at 40 °C to the values measured for TiN.

5. An essential increase in the corrosion resistance was
   observed for the outer titanium nitride zone. Corrosion resistance in the region of the diffusion zone
   continuously decreased with the increasing abrasion
   depth.

6. Changes in corrosion resistance observed in the present study for the nitrided Cr18Ni9Ti steel can be
   related to variations of chromium nitrides precipitation under different nitriding conditions.

REFERENCES

1. B. Edenhofer, HTM, 1974, 29, 105.
2. T. Bell, Z.L. Zhang, J. Lanagan, and A.M. Staines,
   "Coatings and Surface Treatment for Corrosion and
   Wear Resistance", Eds. K.N. Strafford, P.K. Datta,
   and C.G. Googan, Ellis Horwood, Chichester, 1984,
   p.164.
3. P. Süry, Br. Corros. J., 1978, 13, 31.
4. W.A. Mannehimer and H.W. Paxton, Proc. Conf. "Fundamental Aspects of Stress Corrosion Cracking"
   (Columbus, 1967), Eds. R.W. Staehle, A.J. Forty, and
   van D. Rooyen, NACE, Houston, 1969.
5. Z.L. Zhang and T. Bell, Surf. Eng., 1985, 1, 131.
6. J. Flis, J. Mańkowski, and Roliński, Surf. Eng.,
   1989, 5, 151.
7. A.F. Smith and H.E. Evans, J.Iron Steel Inst., 1973,
   34 (Part 1), 211.
8. D.B. Rayaprolu and A. Hendry, Mater. Sci. Technol.,
   1988, 4, 136.
9. B. Billon and Hendry, Surf. Eng., 1985, 1, 125.

10. O.T. Inal, K. Ozbaysal, E.S. Metin, and N.Y.Pehlinvanturk, Proc. ASM's 2nd Intern. Conf. "Ion Nitriding and Ion Carburizing" (Cincinnati, 1989), ASM International, Materials Park, 1990, p.57.

11. K. Ozbaysal and O.T. Inal, J. Mater. Sci., 1986, 21, 4318.

12. E. Roliński, "Plasma Nitriding of Titanium and its Alloys" (in Polish), Warsaw University of Technology Publications, Warsaw, 1988.

13. M. Yan and C. Wang, Proc. 4th Conf "Carbides - Nitrides - Borides", (Poznań - Kołobrzeg, 1987), Poznań Technical University Publishers, Poznań, 1987, p.163.

14. R. Bandy and van D. Rooyen, Corrosion, 1985, 41, 228.

15. S.J. Pawel, E.F. Stansbury, C.D. Lundin, Corrosion, 1989, 45, 125.

16. A. Ramchandani and J.K. Dennis, Heat Treat. Metals, 1988, 15, 34.

17. P.A. Dearnley, A. Namvar, G.G.A. Hibberd, and T. Bell, Proc. 1st Int. Conf. "Plasma Surface Engineering"(Garmisch-Partenkirchen, 1988), Eds. E. Broszeit et al., DGM Informationsgesellschaft, Oberursel, 1989, Vol. 1, p.219.

18. K. Ichii, K. Fujimura, and T. Takase, ibid. Vol. 2, p. 1187.

19. A. Rota, B. Elsener, and H. Böhni, "Surface Engineering Practice", Eds. K.N. Strafford, P.K. Datta, and J.S. Gray, Ellis Horwood, Chichester, 1990, p.587.

20. E.I. Meletis, W.B. Carter, and R.F. Hochman, Microstructural Sci., 1986, 13, 417.

1.2.4
Electrochemical and Electronoptical Investigations of Conversion Coating Growth on Mild Steel

N. Bretherton, S. Turgoose, and G. E. Thompson

CORROSION AND PROTECTION CENTRE, UMIST, PO BOX 88, MANCHESTER M60 1QD, UK

1 INTRODUCTION

The origins of phosphating can be traced to Coslett's[1] original patent. In the employed bath, boiling phosphoric acid containing iron filings, iron and steel articles acquired a protective coating after treating for 2.0-2.5 hours. The development of phosphating has concentrated upon reducing treatment time, to the order of minutes, lowering the bath operating temperature, improving coating performance and widening their application.

The literature gives reviews of phosphating, taking into account theoretical considerations and the many applications [2-5]. The general view is that the mechanism of phosphate coating formation is complex, but it depends on the following equilibrium:

$$Primary Phosphate \rightleftharpoons Tertiary Phosphate$$
$$Soluble \qquad Insoluble$$

Elevated bath temperatures, or a rise in pH, shift the equilibrium to the right. Phosphate film formation relies on two coupled electrochemical reactions. Dissolution of the metal occurs at micro-anodes, or active centres, [2] combined with the discharge of hydrogen at micro-cathodes. Hydrogen evolution leads to a rise in the local pH, resulting in the hydrolysis and precipitation of insoluble phosphates [4]. The presence of zinc iron phosphate (phosphophyllite) in zinc phosphate (hopeite) coatings has been demonstrated by Cheever[6]. This was postulated to arise from an accumulation of ferrous ions at the metal-solution interface in virtually stagnant conditions. Acceleration of coating formation, thereby reducing treatment time, is achieved by the addition of heavy metal salts or oxidizing agents. The action of accelerators has been related to the depolarization of hydrogen reduction [4], with a secondary effect of controlling bath iron content.

The electrochemical nature of coating formation has been used to study surface processes. Although potential-time logging has been utilised [7,8], it provides little information on the involvement of the metal substrate in the coating process. This information can be obtained by monitoring the metal corrosion rate. Although this is impossible using traditional techniques such as polarization resistance measurements, a polarization method employed by Ang[9] provides a route to this information. This has been employed to examine further the various stages of conversion coating formation. Corrosion rate studies are combined with potential-time logging and impedance spectroscopy, together with complementary electronoptical examination and analysis.

2 EXPERIMENTAL

Mild steel flag specimens, of dimension 10x50 mm, were prepared from *Q-panels*. The surface was prepared by abrasion to a 600 grit finish, rinsed and dried in a cool air stream. Specimens were stored in a desiccator prior to use. Immediately prior to coating the specimens were degreased, etched in 20% hydrochloric acid for 150 seconds, rinsed and then dried in a cool air stream.

The phosphating bath [10] was prepared using AR grade reagents, containing 15 g dm^{-3} orthophosphoric acid, 4 g dm^{-3} nitric acid and 6.4 g dm^{-3} zinc oxide. The bath was operated at 85°C. Acceleration was achieved by the addition of 1 g dm^{-3} of sodium chlorate or sodium nitrite. Coating was normally carried out for periods up to 30 minutes duration.

Potential measurements were made with respect to a saturated calomel electrode immersed in a remote reservoir at ambient temperature. Potentiostatic measurements utilized a three electrode system, consisting of a pair of nominally identical mild steel specimens, forming the working and reference electrodes, and a platinum foil counter electrode. This arrangement permits the application of a 10 mV anodic or cathodic polarization to the working electrode with respect to its natural immersion potential, measured by the nominally identical reference electrode. Potential control with respect to a calomel electrode is impossible as rapid changes in the corrosion potential are observed during the early stages of immersion and this results in varying degrees of overpotential. The derivation of corrosion rates assumes that the working and reference electrodes behave similarly and that polarization does not affect film growth. Electrochemical studies, monitoring the potential of both the working and reference electrodes, and subsequent electronoptical examination suggested that this assumption was valid. Potential logging, at a rate of 1 Hz, was achieved using a *PCL-718 Data Acquisition Card* in a Personal Computer using *Labtech Acquire* software.

Impedance spectroscopy was applied to "complete" coatings, after approximately 20 minutes immersion. Both potential and current monitoring indicated that the system had reached an approximate steady state by this time. Spectra were obtained using a *Capcis-March Voltech TFA and Interface* controlled by their *Harmony* software package. Data was analysed using the *Equivalent Circuit* data analysis system developed by Boukamp[11].

3.   RESULTS

Potential-Time Response

Data derived from potential-time studies of coating growth are presented in Figure 1. These show the effect of both chlorate and nitrite acceleration on the corrosion potential shift observed during film formation at 85°C. Coating formation results in a marked increase in potential, greater than 20 mV, from the initial corrosion potential. This is terminated by a potential plateau, where a relatively stable value is achieved. Chlorate acceleration resulted in a slight, 5 mV, decrease in the potential to a stable value after passing through a maximum.

The attainment of a relatively stable potential plateau appears to correlate with the completion of coating formation. Acceleration of the coating bath reduces the time required to achieve this plateau and, hence, the period required for film formation. Potential-time data derived from normal and nitrite accelerated baths are very similar.

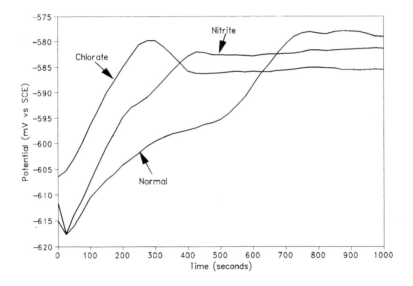

Figure 1 Potential-time behaviour of mild steel immersed in normal and accelerated phosphating baths at 85⁰C

Corrosion Rate Studies

Corrosion rate data derived from both the normal, un-accelerated bath, and the chlorate and nitrite accelerated baths at 85°C are presented in Figure 2. The data are mean values derived from the results of five anodic and cathodic polarization studies. Anodic and cathodic data, i_a and i_c, are obtained by polarization of one mild steel electrode to +10 mV and -10 mV with respect to a second nominally identical reference electrode. The corrosion current density is approximated by application of the Stern-Geary equation. The value of R_p is derived using the following equation, using 10 mV anodic and cathodic polarization:

$$R_p = 0.02 / (i_a + i_c) \tag{1}$$

Anodic and cathodic Tafel constants of 0.04 and 0.12 V, respectively, were used in the estimation of the corrosion rate, producing a value of 0.013 for the constant:

$$i_{corr} = 0.013 / R_p \tag{2}$$

Although this method provides only an approximation of the corrosion rate the rapidly changing corrosion potential precludes more accurate measurement.

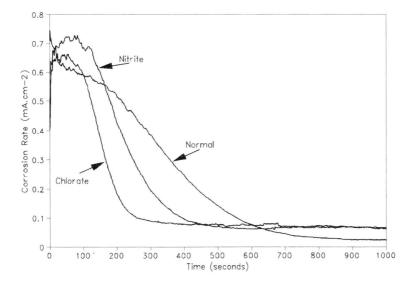

<u>Figure 2</u> Variation of corrosion rate with time during immersion of mild steel electrodes in normal and accelerated phosphating baths at 85°C

Corrosion rate-time data derived from studies at 85°C in the various solutions are all similar in appearance, consisting of a large increase immediately after immersion followed, after a varying period of time, by a decrease to a considerably lower, constant value. These "steady state" values were approximately 2.3×10^{-5} and 6.4×10^{-5} A.cm^{-2} in the normal and accelerated coating baths, respectively.

The results of corrosion rate studies at 85°C correlate with the observed changes in the corrosion potential during coating formation. The large decrease in the corrosion current to a relatively constant value corresponds approximately to the attainment of a relatively stable corrosion potential.   Thus it appears to correlate with coating completion.   It also clearly demonstrates the effect of acceleration, which decreased by more than 50% the time taken to form the complete conversion coating.   The marked difference in the residual corrosion currents suggests a difference in the physical properties of the films formed, possibly the degree of surface coverage and porosity.   Corrosion rate data clearly indicates various stages of coating formation [7].   These can be defined as the initial rise in the corrosion rate to a maximum value, the subsequent onset of a decrease in corrosion rate, the decay of the corrosion rate to a relatively constant low value, and continued growth after coating completion, when a relatively stable residual corrosion rate is achieved. The various stages of film formation were investigated electronoptically.

Electronoptical Investigation of the Various Stages of Coating Formation

Corrosion rate studies identified four stages of coating formation with respect to time of immersion.   Conversion coatings were prepared corresponding to the individual stages and examined electronoptically.   Longer term coating growth was investigated by immersing specimens for a period twice that identified as corresponding to completion.

In both normal and accelerated coating baths, the rise of the corrosion rate to a maximum corresponded to the formation of discrete three-dimensional crystals on the mild steel surface, as is shown in Figure 3.   Crystals were of the order of 100 μm long, leaf-like in structure and grew out from the metal surface.

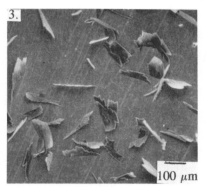

Figure 3 Scanning electron micrograph of mild steel after
immersion for the corrosion stage of film formation

In the normal bath, up to the onset of the corrosion rate decrease, crystal growth continued, leading to approximately 50% surface coverage.   Each crystal appeared to consist of individual layers.   EDX analysis indicating the presence of zinc, phosphorus and iron (probably from the underlying surface), suggested that the crystals were composed of zinc phosphate.   There was no evidence of a film, either visual or analytical, on the surrounding surface as shown in Figure 4a.   Acceleration of the coating process produced a considerably greater degree of surface coverage. Although dendritic crystal growth was visible (Figure 4b) the surrounding surface was also covered with a film. Surface features were still visible through this film, with EDX analysis indicating the presence of traces of phosphorus, and iron, but no zinc. The dendritic crystals contained zinc.

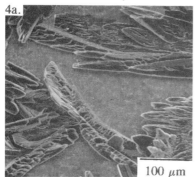
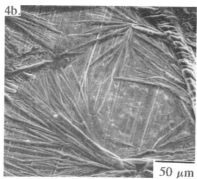

Figure 4 Scanning electron micrographs of mild steel surface after immersion until the initiation of the corrosion rate decrease in various baths: (a) normal coating bath; (b) accelerated coating bath

After immersion times corresponding to coating completion in the normal bath, crystal growth had produced greater than 90% surface coverage. Large areas of apparently "bare" surface were evident (Figure 5a) with EDX analysis detecting only iron at such points. Accelerated coating generated a complete underlying surface film, although dendritic crystal growth was beginning to develop above this film (Figure 5b).

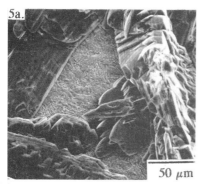
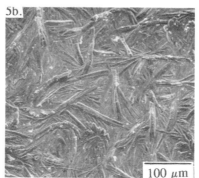

Figure 5 Scanning electron micrographs of mild steel surface after immersion until coating completion in various baths: (a) normal coating bath; (b) accelerated coating bath

The development of coatings for longer than completion in the un-accelerated bath led to the development of a three dimensional crystalline film over the entire surface, with no evidence of "bare" areas. In the accelerated coating bath the underlying surface film was completely covered by dendritic crystal growth. Analysis of both coatings suggested that the visible film was composed mainly of zinc and phosphorus, probably in the form of hopeite (zinc phosphate).

The results of electronoptical examination of the individual stages of coating growth clearly show the effect of accelerators on coating growth. In the normal bath coating formation is via crystal growth, eventually leading to complete surface coverage. This view is supported by the gradual decrease in the corrosion rate to a minimum. Coating completion, identified through corrosion rate measurement, did not appear to correspond to complete surface coverage with "bare" areas still detectable at this point. Acceleration of the coating process initially generates a film covering the majority of the surface. It is suggested that this film is responsible for the initiation of macroscopic surface coverage with a crystalline zinc phosphate and hence contributes to the large decrease in the corrosion rate.

Impedance Spectroscopy

Impedance spectra were obtained after coating completion, approximately 20 minutes after immersion. Typical spectra obtained from a normal coating bath and that from an accelerated coating bath, together with fitted data derived from this spectrum, are presented in Figures 6 and 7. Data derived from both normal and accelerated coatings were of similar appearance, consisting of three overlapping features with frequencies for Z''_{MAX} of approximately 10 kHz, 200 Hz and 630 mHz, and exhibited a significant degree of depression at the high frequency end.

Analysis of the data [11] identified the equivalent circuit shown in Figure 8 as capable of modelling spectra derived from both the normal and accelerated baths. It was not possible to introduce the depression observed at high frequencies to the simulated data; this may be a result of surface roughness or non-uniform current distributions in experimental studies. The high frequency response can be ascribed to the capacitance of a film, C, formed upon the metal surface. Capacitances of the order of 1×10^{-7} F cm^{-2} correspond to an approximate average film thickness of the order of 1×10^{-8} m. Films of this magnitude would be difficult to observe or analyse electronoptically. An alternative explanation could be an instrumental source, producing a high frequency phase shift giving rise to the observed feature at 10 kHz. The low frequency response can be ascribed to a corrosion process, together with an associated diffusional impedance, a finite diffusion length Warburg impedance, O. Derived capacitances were of the order of 10^{-6} F cm^{-2}. The magnitude of the total resistance, total diameter of the three features, is analogous to the polarization resistance derived from DC measurements. Calculating corrosion rates as previously gives values of 4.5 and 6.8×10^{-5} A.cm^{-2} for normal and accelerated coating respectively. These are of similar magnitude to residual corrosion rates observed in DC studies.

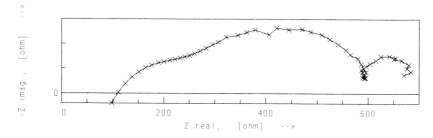

<u>Figure 6</u> Impedance spectra derived from mild steel immersed in the normal phosphating bath at 85⁰C, for 20 minutes

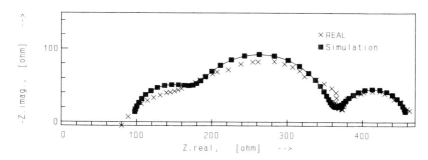

<u>Figure 7</u> Impedance spectra derived from mild steel immersed in the nitrite accelerated phosphating bath at 85⁰C, after coating completion, together with simulated data fitted to the experimental values

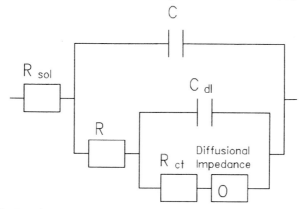

<u>Figure 8</u> Equivalent circuit fitted to experimental data

4 DISCUSSION AND CONCLUSIONS

Both potential-time and corrosion rate studies provide routes for following the development of phosphate conversion coatings, providing a means of detecting apparent coating "completion". Unlike potential monitoring, corrosion rate studies provide an insight into the behaviour of the coatings after the detected end of the film formation process. The significant residual corrosion rates are probably a result of porosity in the coating.

The closely similar maximum corrosion rates observed in normal and accelerated coating formation indicates that chemical accelerators are not involved in the metal corrosion process. Electronoptical studies indicated that, unlike the normal bath, acceleration results in the formation of a surface film immediately prior to the observed rapid decrease in corrosion rate. Available thermodynamic data [12,13] for the formation of phosphate minerals permit calculation of the stability fields for solids formed under these bath conditions at the mild steel surface [8]. The stable solid phases considered are hopeite, $Zn_3(PO_4)_2.4H_2O$, phosphophyllite, $Zn_2Fe(PO_4)_2.4H_2O$, and vivianite, $Fe_3(PO_4)_2.8H_2O$. Data were calculated using a total zinc concentration, $[Zn^{2+}] = 0.079M$ and total phosphate content, $[P_{TOTAL}] = 0.1531M$. Although data were derived for 25°C the general trends should still be relevant at 85°C. The initial bath composition, pH=1.8, indicates that no solid phases are present at 25°C.

Consideration of the stability fields calculated leads to the following conclusions. The formation of hopeite requires a shift of the bath pH. Cathodic evolution of hydrogen is expected to lead to a local increase in pH. Precipitation at cathodic sites has been postulated by Freeman[4] as a mechanism of film formation. This may be a possible route to the formation of an hopeite film. The formation of phosphophyllite or vivianite requires the presence of 0.022 and 0.065M concentrations of ferrous ions in solution at $[Zn^{2+}] = 0.079M$. Corrosion rate data and use of Fick's law, assuming values of the diffusion coefficient and film thickness, allows the calculation of surface ferrous ion concentrations, indicating a maximum surface concentration of 3.5×10^{-3} M. Thermodynamic calculations indicate that this is to low for the formation of phosphophyllite. Since accelerators are chemical oxidants, the oxidation of ferrous may result in the formation of a ferric phosphate film with a calculated extremely low solubility under these conditions. Surface concentrations were calculated using a diffusion coefficient of ferrous in the bulk solution; the formation of this film could reduce these values. This could increase the surface ferrous ion concentration sufficiently to lead to the formation of phosphophyllite at the metal surface. The outer film observed during the later stages of accelerated coating growth is composed of hopeite as no mechanism exists for ferrous ion enrichment at this location.

Impedance spectroscopy derived spectra exhibiting identical features for both normal and accelerated baths after coating completion, suggesting similar surface processes. Both exhibited diffusional impedances at low frequencies, characteristic of diffusion through a film [14]. Consideration of the observed changes in the potential and corrosion rate during coating formation, and of the values of Warburg coefficients suggests that this arises from the diffusion of ferrous ions. Corrosion rates derived from impedance spectra are similar to the values observed in DC studies.

Diffusion coefficients of the order of 1×10^8 cm^2 s^{-1} were derived from the low frequency feature, assuming a film thickness of 1×10^{-6} m. This is considerably lower than values expected in the bulk solution. Surface ferrous concentrations were calculated from parameters derived from the impedance spectra and found to be of the order of 2.25×10^3 M, and are inversely proportional to surface area. If this was restricted, by the porous film, to a fraction of the total surface, the resulting local concentrations would make phosphophyllite formation feasible. Thus, the detection of phosphophyllite in conversion coatings may be a result of its formation after surface coverage with a hopeite film.

In the case of the normal coating bath, film formation appears to result from the growth of discrete crystals of hopeite, with almost complete surface coverage observed at coating completion. The gradual decrease in the corrosion rate supports this model of film growth. The absence of chemical oxidants may account for this marked difference in coating growth, preventing the formation of the initial surface film observed in accelerated baths.

REFERENCES

1.   T.W. Coslett, British Patent no. 8667, 1906.
2.   W. Machu, 'Die Phosphatierung', Verlag Chemie, 1950.
3.   G. Lorin, 'Phosphating of Metals', Finishing Publications Ltd., Middlesex, 1974.
4.   D.B. Freeman, 'Phosphating and metal pre-treatment', Woodhead-Faulkner, Cambridge, 1986.
5.   W. Rausch, 'The Phosphating of Metals', Finishing Publications Ltd., Middlesex, 1990.
6.   G.D. Cheever, J.Paint Tech., 1967, 39, 504.
7.   E.L. Ghali and R.J.A. Potvin, Corros.Sci., 1972, 12, 583.
8.   J.B. Lakeman, D.R. Gabe and M.O.W. Richardson, Trans.IMF, 1977, 55, 47.
9.   K.F. Ang, MSc Dissertation, UMIST, 1989.
10.   B. Zantout and D.R. Gabe, Trans.IMF, 1983, 61, 88.
11.   B.A. Boukamp, 'Equivalent Circuit', Department of Chemical Technology University of Twente, PO Box 217, 7500 AE Enschede, the Netherlands, 1989.
12.   P. Vieillard and Y. Tardy, Thermochemical Properties of Phosphates, 'Phosphate Minerals', eds. J.O. Nriagu and P.B. Moore, Springer-Verlag, Berlin, 1984.
13.   J.O. Nriagu, Formation and Stability of Base Metal Phosphates in Soils and Sediments, 'Phosphate Minerals', eds. J.O. Nriagu and P.B. Moore, Springer-Verlag, Berlin, 1984.
14.   S. Turgoose and R.A. Cottis, Proceedings ASTM Symposium on Electrochemical Impedance Spectroscopy, San Diego, 1991.

1.2.5
The Effect of Chlorinated Hydrocarbons on the Corrosion Resistance of Austenitic Stainless Steels in Chloride Solutions

R. A. S. Hoyle[1] and D. E. Taylor[2]

[1] MARIT METALLURGICAL ENGINEERS LTD, YARM, CLEVELAND, UK

[2] SCHOOL OF TECHNOLOGY, UNIVERSITY OF SUNDERLAND, UK

1   INTRODUCTION

The major problem associated with pitting corrosion is the rapid rate of highly localised attack, leading to failure of a component often by perforation.[1] Pit propagation rates of millimetres per day have been recorded. Austenitic stainless steels may be subject to pitting attack when the normally passive surface film suffers local breakdown. This breakdown may typically result from mechanical damage or chemical dissolution. The lower alloy stainless steels undergo pitting in relatively low concentrations of chloride, higher alloy grades having additions of alloying elements (eg. Cr, Ni, Mo & N) have reduced pitting tendency.[2,3] The grades of austenitic stainless steel used for this work were two of the most widely used in the chemical storage and processing industries, ie. types 304 and 316. Initial investigations were carried out in chloride solutions to determine the nature and behaviour of the pitting attack. Similar tests were also undertaken in the presence of a chlorinated hydrocarbon, 1,2-dichloroethane.[4] Such hydrocarbons are normally immiscible with aqueous solutions and all of the electrochemical investigations took place in the aqueous phases.

    The electrochemical technique used was that of cyclic polarisation,[5] which was employed to study specific variables whilst all other parameters were held as close to constant as practical.

2   PITTING CORROSION

The problems associated with pitting corrosion are well documented. Shreir[6] listed over 250 papers published between 1960 and 1976 dealing with pitting, and at least as many again have been published since that date.

Pit initiation on perfect surfaces, free from any form of heterogeneity, can be explained by two basic theories, kinetic or thermodynamic. The kinetic theory uses the competitive adsorption between oxygen and chloride ions to explain the breakdown of passivity, while the thermodynamic theory considers the equilibrium of the chloride ions and the surface oxide. Both theories have been explained and reviewed by others,[6,7] and are satisfactory for perfect surfaces. However, in practice stainless steels contain inclusions, heterogeneities and second phases, all of which disrupt the perfect surface, and several authors have reported pit initiations at such defects, especially inclusions.[8-11]

The theory of the mechanism behind pit propagation is more widely accepted. It is thought to involve the dissolution of the metal, and the hydrolysis of the metal ions in the bottom of the pit leading to the maintenance of a high degree of acidity. Not all the aspects of this mechanism are fully understood.

The reactions for a stainless steel in neutral aerated sodium chloride solution involve the anodic metal dissolution reactions at the bottom of the pit,

$$Fe \rightarrow Fe^{2+} + 2e^-$$

$$Cr \rightarrow Cr^{3+} + 3e^-$$

balanced by the cathodic reaction on the adjacent surface,

$$O_2 + 2H_2O + 4e^- \rightarrow 4OH^-$$

The increased concentration of Fe^{2+} and Cr^{3+} within the pit results in the migration of chloride ions to maintain neutrality. The metal chlorides formed, $FeCl_2$ and $CrCl_3$, are hydrolysed by water to the hydroxides and free acid,

$$FeCl_2 + 2H_2O \rightarrow Fe(OH)_2 + 2HCl$$

$$CrCl_3 + 3H_2O \rightarrow Cr(OH)_3 + 3HCl$$

The generation of this acid lowers the pH values at the bottom of the pit to as low as 1.3[7] while the pH of the bulk of the solution remains neutral.

Temperature was considered as a pitting criterion in 1973[12] when it was suggested that there exists for each wholly austenitic stainless steel a critical pitting temperature, below which the steel will not pit regardless of potential and exposure time. An empirical equation for the calculation of the critical pitting temperature for the average pitting behaviour of all the steels tested in the range 3 to 7.5% molybdenum, was put forward:

Critical Pitting Temperature($^{\circ}$C) = 10 + 7(%Mo).

However it was found that the temperature values
obtained by use of this equation were only valid for flat
surfaces, and the more likely cases of edge or crevice
attack were liable to occur at much lower temperatures.

3   CYCLIC POLARISATION PRINCIPLE AND PRACTICE

Conditions within the cell are stabilised and measurement
made of the free corrosion potential, or rest potential,
E_{rest}. This is the potential to which the sample will
naturally return if left without impressed voltages. The
cyclic polarisation curve is produced by commencing a
potential sweep at a value of potential below the
corrosion equilibrium point, E_{corr}, and sweeping in a
noble direction to some predetermined point above the
pitting potential, E_p. This point may be determined
either from a potential or current value, by either having
a maximum potential or current at which the sweep
direction is reversed. The reversed sweep curve should
return to cross the upward part of the curve at a point
known as the repassivation, or protection potential,
E_{pass}. The value of rest potential, E_{rest}, may be either
above or below the value of the repassivation potential,
E_{pass}, see Figure 1.

The pitting potential, E_p, when measured by this
method is the point at which the logarithm of current
density begins to increase rapidly with increasing
potential.[13]

The general trend is that the more positive the
pitting potential, the less the likelihood of pit
initiation, and the more positive the repassivation
potential, the less the likelihood of pit propagation.

Figure 1 shows a schematic cyclic polarisation curve,
illustrating conditions under which pitting may occur.[1]

4   THE USE AND ANALYSIS OF ELECTROCHEMICAL DATA

A number of variations of technique in the use of
electrochemical polarisation measurements, for the
determination of resistance to localised corrosion, have
developed over the last 30 years. Of these, probably the
most relevant to the present study is that published by
the American Society for Testing and Materials, ASTM G61.
The standard includes representative cyclic polarisation
curves and standard polarisation plots, for both the
forward and reverse scans, giving acceptable limits for
the current density values obtained during the scan.

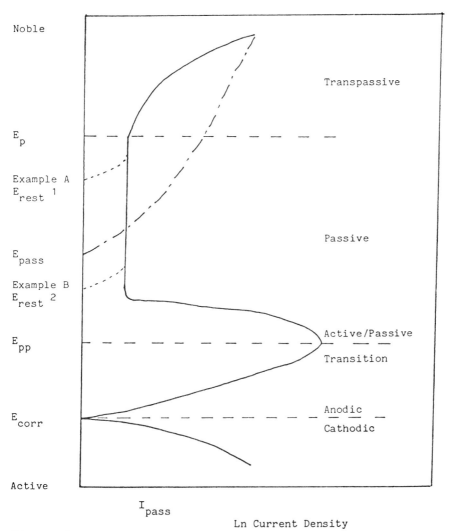

Figure 1

Example A. $E_{rest} > E_{pass}$ Pitting will continue once initiated.

Example B. $E_{pass} > E_{rest}$ Repassivation of pit sites will occur.

5   EXPERIMENTAL PROGRAMME

Test Equipment

Equipment for this investigation included a Chemical Electronics Company potentiostat, type TR40-3A, driven by a linear sweep generator. A three electrode system was connected to the potentiostat and a pair of digital multimeters used to give constant visual monitoring of the current and potential in the cell.

The test cell was a 1 litre round bottomed glass flask, supported in an electric mantle, with a 'quickfit' lid giving five access points to the cell. Three of these were used for the working (stainless steel), reference (saturated calomel) and secondary (platinum) electrodes, the remaining two being used for a thermometer and air supply, if and when required.

Materials

Two grades of austenitic stainless steel were used, both supplied in the form of sheet. The grades used investigated BS1449 pt2 316S16 (type 316) and BS1449 pt2 304S15 (type 304), analysis and history of which are shown in Table 1. Both materials had been cold rolled, softened and descaled, and the type 316 stainless steel had been pinch passed.

As can be seen from Table 1, the significant compositional differences between the two grades of austenitic stainless steel chosen for this work were in the chromium, molybdenum and nickel contents.

Samples

The working electrode was designed to expose a known area of metal to the test environment, and be suitable for reconditioning and reuse.

A 10mm x 10mm square was cold cut from sheet, the edges dressed to p1200 grade on emery, a tolerance of 0.05mm being allowed on any single 10mm edge. The samples were attached to a length of pvc or silicone rubber insulated wire and cast into a polyester block, exposing one face of the sample. The insulated wire exited the block from either the side or top, depending on the orientation required for the sample. The working face of the sample was dressed to the appropriate grade of emery, or polished on diamond impregnated cloths using conventional metallographic practice. Several such

<u>Table 1</u> Analysis and history of the test materials

Element Wt%	Material Type 316	Type 304
Carbon	0.038	0.05
Silicon	0.28	0.37
Manganese	1.67	1.49
Phosphorus	0.030	0.023
Sulphur	0.010	0.005
Chromium	16.70	18.24
Molybdenum	2.30	-
Nickel	11.60	8.69
Cast Number	P9864	W4193
Supplier	BSC Stainless	BSC Stainless
History	Cold Rolled, Softened, Descaled & Pinch Passed.	Cold Rolled, Softened & Descaled.

samples were prepared from each grade of stainless steel, but all the material of a specific grade was taken from the same sheet.

<u>Electrolytes</u>

The solutions used throughout the test programme are listed below with a brief description.

NaCl   A solution of AnalaR grade sodium chloride in distilled water giving a concentration of 3.56% by weight sodium chloride.

NSW   Fresh North Sea Water, collected within the preceding 24 hours, from the Blackhall Rocks area, north of Hartlepool, Cleveland.

EDC   Commercially pure 1,2-dichloroethane.

Distilled   Distilled fresh water.

NaCl+EDC   A solution of 3.56% by weight sodium chloride with the addition of 50ml of 1,2-dichloroethane per 1000ml aqueous solution.

NSW+EDC   A solution of fresh North Sea Water with an addition of 50ml of 1,2-dichloroethane per 1000ml aqueous solution.

Distilled+EDC   A solution of distilled fresh water with the addition of 50ml of 1,2-dichloroethane per 1000ml aqueous solution.

6   EXPERIMENTAL PROCEDURE

Experimental procedure varied depending on the parameter being investigated; however, in all cases the initial preparation of the test specimens was the same.

The selected test electrode was prepared not more than one hour prior to the commencement of the test, by grinding or polishing to the required grade of emery, using normal metallographic preparation techniques.

In a typical test, the appropriate solution was placed in the cell and the three electrodes, thermometer and air diffuser inserted if required. The cell was then heated and stabilised at the selected test temperature. If air was required it was turned on prior to the heating cycle, and left on throughout the test. The temperature having stabilised, the sample was cathodically cleaned, if required, by holding at -1.5V(SCE) for 10 minutes. The cell was then isolated for a period of up to 30 minutes to allow the potential to stabilise and the rest potential determined. When the change of potential was less than 1mV per minute the value observed was taken as being the rest potential.

In most cases the cyclic sweep commenced at the rest potential and swept up to +1V(SCE) when it reversed automatically and swept back down to the potential at which zero current was observed.

In some cases the cyclic sweep was started at -1V(SCE) and swept to +1V(SCE) so as to establish the shape of the full cyclic polarisation curve, and allow an estimate the location of the rest potential. Alternatively, the sweep was reversed at +1V(SCE) and returned to -1V(SCE).

Following each test the stainless steel test electrode was rinsed and visually examined, microscopic examination being carried out if appropriate.

7   TEST PROGRAMME

The experimental test programme consisted of a series of cyclic polarisation tests in which selected variables were investigated and the three potentials of rest, pitting and repassivation determined. The test programme centred on the variables listed below.

        i.   Cathodic cleaning,
        ii.   Orientation,
        iii. Solution composition,
        iv.   Surface finish,
        v.   Temperature.

Results

Cyclic Curves. Of the three types of curve plotted the latter type produced by far the most useful data as it indicated not only the pitting potential and the repassivation potential, but the region of the curve in which the rest potential lies.

Consistency of Results. Typical variations of pitting potential, repassivation potential and rest potential were +/-50mV, or twice the recorded voltage increment used, over each set of tests under similar conditions.

Interpretation. This involved analysis of the shapes of the curves to establish the points at which the transition from passive to active state occurred, ie. the pitting potential, and the point at which the return curve reached and crossed the initial curve, ie. the repassivation potential.

In all cases where a rest potential and repassivation potential was recorded, it was possible to establish whether the sample would, in the case of an initiation site being formed, continue to pit, or repassivate. The formation of such an initiation site could result from mechanical damage or the momentary breakdown of the passive surface film due to chemical action.

Where both the pitting and repassivation potentials were measured, that for pitting potential was significantly the more anodic indicating that the likelihood of pit propagation following initiation would be relatively high. In those cases where the rest potential was also appreciably more anodic than the repassivation potential, the risk of serious pitting corrosion taking place would be regarded as very high.

Graphs of the trends in the results which were observed when specific variables were changed were produced. Each of the variables had a significant effect on the three potentials being monitored. The least clear effect was temperature. Although a general trend could be observed for this variable, the trend over a short temperature span did not necessarily agree with that over the full range tested.

8   DISCUSSION

Trends in the Results

Graphs of the trends observed, when specific variables were changed, were produced and are discussed below.

Cathodic Cleaning. The cathodic cleaning acts to remove oxides and other contaminants from the surface of the sample, and this results in the true metal surface being exposed. When the rest potential was measured it was more cathodic due to the absence of this oxide film.

Table 2 identifies a reduction in the pitting and rest potentials, but an increase in the repassivation potential from marginally below the rest potential to significantly above it following cathodic cleaning. This change in potentials would prevent pit propagation at the rest potential of the cathodically cleaned sample, whereas the sample which was not cathodically cleaned could continue to propagate pits at the rest potential.

Table 2 Effect of cathodic cleaning on Type 304
stainless steel in sodium chloride solution

Condition	Rest Pot. mV	Pitting Pot. mV	Repassivation Pot. mV
Not Cath. Cleaned	-228	187	-243
Cath. Cleaned	-356	120	-167

Table 3 Effect of orientation on Type 316 stainless
steel in sodium chloride solution

Orientation	Rest Pot. mV	Pitting Pot. mV	Repassivation Pot. mV
Vertical	-265	275	-205
Horizontal	-139	225	-220

The fall in pitting potentials is also explained in terms of the oxide film on the surface of the metal. During cathodic cleaning the oxide film is destroyed and it subsequently reforms following the removal of the cathodic cleaning potential. This results in a new, uniform but weak film being present when the test is commenced. As a consequence of this the film is more readily broken down and pit initiation occurs more easily, and hence at lower potentials.

Sample Orientation. Tables 3 & 4 illustrate the effects of orientation on both type 316 and type 304 stainless steels.

Table 3 shows that there is a slight reduction of pitting potential and of repassivation potential, but a significant increase in rest potential when switching from the vertical to horizontal orientation.

Table 4 indicates a very slight reduction of pitting potential and a slight increase of rest potential, but no significant change in repassivation potential, which remained below the rest potential for both orientations. All these variations were within experimental error.

The pitting potential increased for both steels when switching from vertical to horizontal orientations because the corrosion products formed when a pit initiates but fails to propagate remain on the sample surface. This leads to easier pit initiation and a second pit may initiate at the same site. When the sample is vertically orientated, the corrosion product falls from the surface and is of no help in further pit initiations. The repassivation potential was noted to fall slightly, or not change, and this is due to a similar reason. The corrosion products formed in the pits act to increase the aggressive nature of the solution in the pit. If the

Table 4 Effect of orientation on Type 304 stainless steel in sodium chloride solution

Orientation	Rest Pot. mV	Pitting Pot. mV	Repassivation Pot. mV
Vertical	-228	187	-243
Horizontal	-203	175	-250

Table 5 Effects of solution on Type 316 stainless steel

Solution	Rest Pot. mV	Pitting Pot. mV	Repassivation Pot. mV
North Sea Water	-117	268	-218
Sodium Chloride Solution	-265	275	-205

sample is orientated horizontally, these corrosion products will remain in the pit, whereas if the sample is orientated vertically, some of these corrosion products will fall out and be replaced by fresh solution. Hence the repassivation potential may be decreased when these aggressive corrosion products remain in the pits. The effect is very small.

The rest potential may have increased in the horizontal orientation due to surface contamination from particulate matter within the solution depositing on the sample and thus preventing the formation of the protective oxide film, or disrupting the film already formed.

Solution Composition. Tables 5, 6 and 7 indicate the effects of solution composition on both type 316 and type 304 stainless steels. Solutions tested were natural North Sea Water compared to sodium chloride solution and to distilled water.

Table 5, a comparison of North Sea Water and sodium chloride solution using type 316 stainless steel, shows a large reduction of rest potential, but a slight increase of both pitting potential and repassivation potential when changing from North Sea Water to sodium chloride solution.

Table 6, again a comparison of North Sea Water and sodium chloride solution, but using type 304 stainless steel, shows a large increase of rest potential and of repassivation potential, but a reduction in pitting potential. The rest potential increase is just large enough to result in the repassivation potential being below the rest potential for the sodium chloride solution.

Table 6 Effect of solution on Type 304 stainless steel

Solution	Rest Pot. mV	Pitting Pot. mV	Repassivation Pot. mV
North Sea Water	-320	275	-300
Sodium Chloride Solution	-228	187	-243

Table 7 Effect of solution on Type 316 stainless steel at 70°C

Solution	Rest Pot. mV	Pitting Pot. mV	Repassivation Pot. mV
North Sea Water	-350	-175	-275
Distilled Water	-148	-	-175

Table 7 indicates the differences between North Sea Water and distilled water on type 316 stainless steel at 70degC. In distilled water the pitting potential could not be established. In the North Sea Water the repassivation potential was above the rest potential, but in the distilled water, while both potentials were higher, the rest potential was slightly above the repassivation potential.

The difference in effects between North Sea Water and sodium chloride solution on type 316 stainless steel may result from North Sea Water not being a simple solution, but having many constituants, and being acidic. This acidity has the effect of reducing the ability of the stainless steel to resist corrosion, with the result that the rest potential is raised and the pitting and repassivation potentials are lowered.

For type 304 stainless steel the apparent effects are that the sodium chloride solution is more corrosive than the North Sea Water. The reason for this is not clear and further study is being undertaken.

Surface Finish. Tables 8 and 9 show the effects of surface finish on the materials under test. Finishes of 1um diamond polish, p1200 and p200 grade wet emery grind were compared.

Table 8 indicates the effect of surface finish on type 316 stainless steel. It shows that the pitting potential is dependent on the quality of polish: the better the polish the higher the pitting potential. The effect on repassivation potential is similar but not so

Table 8 Effects of surface finish on Type 316 stainless
        steel in sodium chloride solution

Surface Finish	Rest Pot. mV	Pitting Pot. mV	Repassivation Pot. mV
1um Polish	-337	375	-60
p1200 Emery	-256	275	-205
p220 Emery	-344	125	-210

Table 9 Effects of surface finish on Type 304 stainless
        steel in sodium chloride solution

Surface Finish	Rest Pot. mV	Pitting Pot. mV	Repassivation Pot. mV
1um Polish	-199	250	-150
p1200 Emery	-228	187	-243
P220 Emery	-197	275	-180

extreme, with the repassivation potential reducing from the value of the 1um to that of the p1200 grade but then levelling out when switching to the p220 grade. The rest potential is lower for the very fine 1um finish and at the coarser p220 grade, but high at the intermediae p1200 grade.

The effect of surface finish on type 304 stainless steel is shown in Table 9. All three potentials are lower for the p1200 grade than for either of the others, with the 1um and p220 grade finishes producing very similar results.

For type 316 stainless steel the rest potential was seen to rise when changing from the 1um to the p1200 grade, and then fall when changing to the p220 grade. This was attributed to the high quality polish of the 1um the surface. The rest potential is lower in the 1um case because the ease of oxide film formation is increased, there being fewer facets to interupt the film growth. On the p1200 grade samples the surface is roughened and the oxide film has to grow on several differing planes, thus slowing the overall film formation and hence the film strength. The same argument can be applied to the effect between p1200 grade and p220 grade. Although the rest potential falls from the p1200 grade to the p220 grade, the relative positions between the rest potentials and the pitting potentials is narrowed. This indicates that the protection offered by the oxide film on the p220 grade samples is less than that offered by the film on the p1200 grade samples.

The fall of pitting potential for type 316 stainless steel with coarser finishes is also due to the quality of the surface finish. The possible sites at which pits can initiate are few on the 1um sample, and as such pit initiation can be expected to be delayed when compared to the p1200 grade samples. This would logically suggest that pitting would be more easily initiated on the p220 grade sample, and this is supported by the fact that the pitting potential on the p220 grade sample was indeed below that on the p1200 grade sample.

The repassivation potential of type 316 stainless steel reduced when changing from the 1um to the p1200 grade due to oxide film growth becoming more difficult, for the reasons detailed above, but the fall when changing from the p1200 grade to the p220 grade was much less as the p1200 grade is already sufficiently roughened to require a much larger film initiation network, and further roughening has little effect. This case is slightly different to that detailed above for the rest potential, as the pits formed in the solution during the test had no oportunity to initiate oxide films prior to the repassivation potential.

For type 304 stainless steel a similar situation arises, but the extent to which it may be observed is much diminished. The fact that the rest potential is above the repassivation potential for the p1200 grade, and then below it for the p220 grade indicates how unstable the conditions are for this material.

Solution Temperature. Tables 10 and 11 show the
effects of solution temperature on type 316 and type 304
stainless steels respectively.

A steady fall in pitting potential is shown, in
Table 10 for type 316 stainless steel and Table 11 for
type 304 stainless steel, as temperature increases. The
rest potentials also show a slight reduction, though the
trend is not very clear. The repassivation potential,
however, indicates a clear trend of reducing values as the
temperature increases.

As temperature increases, the ease with which the
protective oxide film on both type 316 and type 304
stainless steel breaks down increases and hence the
pitting potential falls. The slight falls in both the
rest and repassivation potentials are due to the higher
energy state of the solution, and hence the higher
mobility of species present. This increases the speed at

Table 10 Effects of temperature on Type 316 stainless
         steel in sodium chloride solution

Temperature	Rest Pot. mV	Pitting Pot. mV	Repassivation Pot. mV
Ambient	-265	275	-205
40°C	-227	75	-330
50°C	-368	50	-210
60°C	-186	-50	-290
70°C	-247	25	-270
80°C	-258	-150	-540
90°C	-323	-50	-355

which the oxide film can be generated or reformed, but
also the aggressiveness of the chloride ions in the
solution. The result is a general fall in all three
measured potentials, but also a narrowing of the gap
between the pitting potential and the other two
potentials.

Addition of 1,2-Dichloroethane

The effect of the addition of 1,2-dichloroethane
(EDC) to sodium chloride solution on type 316 stainless
steel is shown in Table 12. The rest potential increased
and the repassivation potential decreased slightly, but
remained just above the rest potential. The pitting
potential also decreased when EDC was added.

Table 11 Effects of temperature on Type 304 stainless
steel in sodium chloride solution

Temperature	Rest Pot. mV	Pitting Pot. mV	Repassivation Pot. mV
Ambient	-228	187	-243
40°C	-210	100	-290
50°C	-245	-50	-265
60°C	-298	-25	-360
70°C	-283	-25	-280
80°C	-307	-50	-300
90°C	-236	-130	-355

Table 13 also shows the effect of the addition of EDC
to sodium chloride solution, but on type 304 stainless
steel. Tests on pure EDC are also shown. Unfortunately,
only rest potential could be measured in the pure organic
liquid. However, when EDC was added to sodium chloride
solution the rest potential and pitting potential
increased significantly, and the repassivation potential
decreased to be almost coincident with the rest potential.
The rest potential in pure EDC lay between those for the
sodium chloride solution and that with EDC added.

Table 12 Effect of addition of EDC to sodium chloride
solution on Type 316 stainless steel

Solution	Rest Pot. mV	Pitting Pot. mV	Repassivation Pot. mV
Sodium Chloride Solution	-265	275	-205
NaCl + EDC	-234	237	-210

Table 13 Effect of addition of EDC to sodium chloride
solution on Type 304 stainless steel

Solution	Rest Pot. mV	Pitting Pot. mV	Repassivation Pot. mV
Sodium Chloride Solution	-345	-57	-225
NaCl + EDC	-238	33	-233
EDC	-300	-	-

The effect of the addition of EDC to North Sea Water is shown in Table 14. The pitting potential of type 304 stainless steel increased, while the rest potential decreased appreciably and the repassivation potential increased significantly.

Table 15 shows the effects of adding EDC to distilled water, and of pure EDC, on type 316 stainless steel at 70°C. Unfortunately, rest potentials only could be measured when EDC was present, and no value of pitting potential could be established for the distilled water. The rest potentials decreased as EDC was added.

When EDC was added to sodium chloride solution the effects on type 316 stainless steel were generally slight. The three potentials moved closer together, indicating a reduction in corrosion resistance, but the changes were not great. Effects on type 304 stainless steel were more significant. The increase in rest potential and pitting potential show generally more passive conditions existed, but the reduction in repassivation potential indicates a condition exists where the passive oxide film is less readily reformed, and the metal is less stable.

When EDC was added to North Sea Water the effects were the same as when adding EDC to the sodium chloride solution.

The remaining tests in the group were all conducted using EDC or distilled water for at least one point on the graph. It was not possible, using cyclic polarisation

Table 14 Effect of addition of EDC to North Sea Water
        on Type 304 stainless steel

Solution	Rest Pot. mV	Pitting Pot. mV	Repassivation Pot. mV
North Sea Water	-350	-175	-275
NSW + EDC	-688	100	-150

techniques, to establish pitting potentials for either of these solutions, or establish repassivation potentials for the EDC.

Possible Mechanism

When EDC was present, an effect was observed on the rest potential, pitting potential and repassivation potential of the test sample. The reason for this effect is thought to be the hydrolysis or dissociation of the EDC to produce HCl.

$$\text{Cl-}\underset{\underset{\text{H}}{|}}{\overset{\overset{\text{H}}{|}}{\text{C}}}\text{-}\underset{\underset{\text{H}}{|}}{\overset{\overset{\text{H}}{|}}{\text{C}}}\text{-Cl} \quad \xrightarrow{\text{H}_2\text{O}} \quad \text{Cl-}\overset{\overset{\text{H}}{|}}{\text{C}}\text{=}\underset{\underset{\text{H}}{|}}{\text{C}} \quad + \text{ HCl} \quad \text{or} \quad \text{Cl-}\underset{\underset{\text{H}}{|}}{\overset{\overset{\text{H}}{|}}{\text{C}}}\text{-}\underset{\underset{\text{H}}{|}}{\overset{\overset{\text{H}}{|}}{\text{C}}}\text{-OH} \quad + \text{ HCl}$$

The production of HCl in this way would be consistent with the observations made of the rest, pitting and repassivation potentials. The presence of HCl in the test cell has not been proven and further work on this aspect is required.

Table 15 Effect of addition of EDC to distilled water on
        Type 316 stainless steel at 70^0C

Solution	Rest Pot. mV	Pitting Pot. mV	Repassivation Pot. mV
Distilled Water	−148	−	−175
Distilled + EDC	−240	−	−
EDC	−350	−	−

9   CONCLUSIONS

i).   Cathodic cleaning causes an increase in the pitting resistance of the stainless steel.

ii). A vertically orientated sample is less susceptible to pitting attack than a horizontally mounted sample.

iii). Natural North Sea Water is more corrosive to the stainless steel than the sodium chloride solution.

iv). A more highly polished surface has a greater pitting resistance than a less polished surface.

v).   Increasing temperature acts to reduce pitting resistance for both materials tested.

vi). The presence of 1,2-dichloroethane in either North Sea Water on the sodium chloride solution acted to reduce pitting resistance of type 316 stainless steel.

REFERENCES

1. A J Sedriks, Corrosion of Stainless Steels,
    Wiley-Interscience, New York, 1979.

2. E M Horn, D Kuron & H Grafen, Werkstoffe und
    Korrosion, Vol.30, pp723, 1979.

3. A J Sedriks, Stainless Steel '84, Proceedings of
    Goteborg Conference, Book No. 320, pp125, Institute
    of Metals, 1985.

4. J J Demo, Corrosion of Metals in Chlorinated
    Solvents, Corrosion, Vol.24, pp139, 1968.

5. J M West, Applications of Potentiostats in Corrosion
    Studies, British Corrosion Journal, Vol.5, pp65,
    1970.

6. L L Shreir, Corrosion, Metal/Environment Reactions,
    L L Shreir, Editor, Newness Butterworths, Boston,
    Mass., Vol.1, pp182, 1976.

7. S Szklarska-Smialowska, Localised Corrosion, NACE,
    Houston, Texas, pp312, 1974.

8. H H Uhlig, Trans. AIMME, Vol.140, pp411, 1940.

9. M A Streicher, J. Electrochem. Soc., Vol.103, pp375,
    1956.

10. B E Wilde & J S Armijo, Corrosion, Vol.23, pp208,
    1967.

11. S Steinmann, Mem. Sci. Rev. Met., Vol.65, pp615,
    1969.

12. R J Brigham & E W Tozer, Corrosion, Vol.29, pp33,
    1973.

13. Metals Handbook (9th Edition), Vol.13, pp271.

1.2.6

Surfactants as Phosphating Additives – Some Experiences

T. S. N. Sankaranarayanan and M. Subbaiyan

DEPARTMENT OF ANALYTICAL CHEMISTRY, UNIVERSITY OF MADRAS, GUINDY CAMPUS, MADRAS 600 025, INDIA

1 INTRODUCTION

Phosphating is the most widely used metal pretreatment process for the surface treatment and finishing of ferrous and non-ferrous metals. Due to its economy, speed of operation and ability to afford excellent corrosion resistance, wear resistance, adhesion and lubricative properties, it plays a significant role in the automobile, process and appliance industries. Though the process was initially developed as a simple method of preventing corrosion, [1,2] the changing end uses of phosphated articles have forced the modification of the existing processes and development of innovative methods to substitute the conventional ones. [3] To keep pace with the rapid changing need of the finishing systems, numerous modifications have been put forth in their development - both in the processing sequence as well as in the phosphating formulations. A review of the literature gives a clear picture of how these developments have been made and under what grounds; the majority of them involve the introduction of special additives in the phosphating bath. The common aim of these modifications is to obtain good, compact, fine-grained, adherent coatings of excellent corrosion resistance in the most economically and environmentally viable method and to reduce the existing multistage process into a single stage process. The present investigation intends to evaluate the role of surfactants, the versatile additives, most commonly incorporated in a phosphating bath.

Surfactants in Phosphating Processes

Literature reports the use of surfactant in corrosion resistant phosphate treatment of metals.[4] Several phosphating baths have been developed with surfactants as one of the useful additives.[5,6] In these baths, the surfactants afford a more rapid action as well as a better and more uniform bond between the coating and the underlying metal. The effect of cationic, anionic and nonionic surfactants have been studied earlier.[6] However, there is a little or no theory to guide the choice of individual surfactants and it is usually empirical. Moreover, most of the formulations are patented and the exact nature and the role of surfactants in the phosphating bath have not been disclosed in detail for the benefit of the users. The present study which intends to evaluate the role of octadecylamine (ODA), octadecyldithiocarbamate (ODDTC) and Span 60, the representative compounds chosen under the cationic, anionic and nonionic categories respectively, focuses mainly on their chemistry and their influence on phosphate coating formation which are of utmost importance in the selection of a proper surface active agent from an industrial point of view.

2 EXPERIMENTAL

Bath Composition and Operating Conditions

A cold zinc phosphating bath was formulated using only the basic chemicals such as zinc oxide, phosphoric acid and sodium nitrite to evaluate the role of the proposed additives without any interference (Table 1). Hot rolled mild steel panels (Composition conforming to IS: 1079 specifications) of size 8cm x 6cm x 0.2 cm were used for coating operations. Phosphating was done by an immersion process at room temperature (27°C) for 30 minutes. The processing sequence adopted is described elsewhere.[7,8]

Methods of Evaluation

The phosphate coatings obtained with and without the proposed additives were studied for the amount of coating formed and the amount of iron dissolved during phosphating. Based on the optimum concentration chosen for each additive, they were assessed for uniformity, coverage, hygroscopicity, absorption value, loss in weight after 24 hours in 3% NaCl solution, performance in the humidity test after 168 hours of exposure and the spreading of corrosion in the salt spray test after 200 hours.

Table 1   **Process Variables and Conditions**

Chemical Composition of the Formulated bath

ZnO	5 g l^{-1}
H_3PO_4	11.3 ml l^{-1}
$NaNO_2$	2 g l^{-1}

Control Parameters

pH	2.71
Free acid value (FA)	3 points
Total acid value (TA)	25 points
FA : TA	1 : 8.33

Operating Conditions

Temperature	Room Temperature (27°C)
Time	30 minutes

3 RESULTS AND DISCUSSION

The Role of ODA

The effect of ODA was studied in the phosphating bath. The addition of various concentrations of ODA (10-125 mg l^{-1}) was found to decrease the coating weight and the amount of iron dissolved during phosphating (Table 2). Though there is a decrease in the amount of coating formed, they are highly uniform and free of defects. The concentration at which maximum coating weight was obtained (25 mg l^{-1}) was choosen as the optimum concentration for further evaluation. The physical properties and corrosion performance of the phosphate coating studied (Tables 3 and 4) reveal that the performance of the coatings obtained in presence of ODA are not inferior to those obtained using the standard bath (blank) inspite of the reduction in coating weight, due to incorporation of this amine additive in the phosphate coating.[8] The observed results strongly support the existing view that the performance of the phosphate coating is largely determined by the uniformity and coverage and not by the coating weight. The additional advantage gained besides the above is that there is a considerable reduction in the chemical consumption of the bath used which implies improved service life of the bath. However, the criteria for the choice of the optimum concentration and the type of baths in which this additive could yield beneficial effects, should not be over simplified. This can be illustrated by the fact that the 16% reduction in coating weight observed in the present case could be detrimental if used in a bath formulated to produce low coating weights where such an additive would further lower the coating weight to an extent that would affect its uniformity and coverage. This leads us to conclude that this additive will be most suited for phosphating baths possessing high acidity where controlled etching is demanded, a situation commonly observed in chromate phosphating baths used for pretreating aluminium. This was substantiated by the results obtained from the use of a similar kind of additive namely hexadecylamine(HDA) in a chromate phosphate bath formulated for this purpose.[9] The study indicates the ability of HDA to reduce the coating weight and to cause a controlled etching of the base metal besides assiting the conversion of Cr(VI) to Cr(III), essential for the deposition of the chromate phosphate coating on aluminium.

The Role of ODDTC

The anionic surface active derivative of ODA, namely octadecyldithiocarbamate(ODDTC), was synthesised as reported earlier[10] and used as an additive in the same standard bath. The addition of this compound (10-125 mg l^{-1}) increases the coating weight and the weight of iron dissolved during phosphating (Table 2); these effects being consequent to its proven complexing ability.[11] As in the case of ODA, the concentration at which maximum coating weight and uniform surface coverage were obtained (50 mg l^{-1}) was chosen as the optimum concentration for further evaluation. As expected, there is a significant improvement in the properties and performance of the coatings obtained in its

Table 2 **Effect of the Additives on Coating Weight and the Amount of Iron Dissolved During Phosphating**

Additive Used	Concentration of the Additive $(mg\ l^{-1})$	Coating Weight* $(g\ m^{-2})$	Weight* of Iron Dissolved During Phosphating $(g\ m^{-2})$
ODA	0	9.92	5.23
	10	7.14	3.98
	25	8.36	4.90
	50	8.06	4.62
	75	7.82	4.48
	100	7.64	4.34
	125	7.40	4.14
ODDTC	0	9.92	5.23
	10	10.48	5.36
	25	11.92	5.60
	50	14.46	6.01
	75	11.69	5.58
	100	10.73	5.40
	125	10.10	5.28
Span 60	0	9.92	5.23
	50	6.67	4.48
	100	7.19	5.21
	500	7.50	5.41
	1000	8.64	6.46
	1500	7.29	5.42
	2000	6.17	4.69

* Average of five determinations.

presence (Tables 3 and 4). In order to compare the performance of the coatings having similar coating weight (as normally practised) obtained using the standard bath and ODDTC containing bath, the processing time was restricted to 18 min. (instead of 30 min.) in the latter case. Since the performance of the coatings obtained in the above two cases were highly comparable, it is concluded that ODDTC can be effectively used to reduce the processing time considerably. Such a reduction in processing time helps in increasing the area of the metal that can be effectively processed per unit time - a feature which has a tremendous impact on the economy of the phosphating industry. Though ODDTC increases the coating weight, it is less suited as an additive to be incorporated in phosphating baths operating at higher pH values, as in such baths the attainment of the point of incipient precipitation is rapid, causing an excessive deposition of the coating having a powdery nature besides producing excessive sludge.

<u>Table 3</u>   **Properties of the Phosphate Coatings Studied**

Additive Used	Colour	Uniformity	Hygroscopicity (% of Coating Weight)	Absorption Value (g m^{-2})
BLANK	Gray	Uniform	0.040	16.23
ODA	Grayish-white	Uniform	0.029	13.35
ODDTC	Gray	Uniform	0.026	10.84
Span 60	Gray	Uniform	0.036	15.86

<u>Table 4</u>   **Corrosion Performance of the Phosphate Coatings Studied**

Additive Used	Immersion in 3% NaCl	Humidity Test	Salt Spray Test
	Weight loss after 24 hours (g m^{-2})	Observations after 168 hours	Ratings[*] after 200 hours
BLANK	13.12	Remains good	4
ODA	11.37	Remains good	2
ODDTC	9.79	Remains good	2
SPAN 60	10.12	Remains good	2

* After ASTM B 117 - 87.

The Role of Span 60

The sorbitol ester of the C_{18} acid namely, sorbitan monostearate (Span 60) was chosen as the additive in the nonionic category for effective comparison. The addition of Span 60 (50 - 2000 mg l^{-1}) produces a similar effect of decreasing the coating weight and the amount of iron dissolved during phosphating as in the case of ODA (Table 2). Using similar criteria as in the earlier cases, an optimum concentration of 1000 mg.l^{-1} was chosen in the case of Span 60. The properties and performance of panels coated in its presence were comparable with that of the standard bath (Tables 3 and 4). Hence it appears that Span 60 would be as effective as ODA when used in baths performing under similar conditions. However, these effects cannot be generalised and the effectiveness of each additive can only be predicted based on its chemical nature in the phosphating bath.

Chemistry of the Additives in the Phosphating Bath

The phosphating bath used in the present study is highly acidic in nature (pH = 2.71). The added ODA will exist as onium cation[12] (R-$\overset{+}{N}H_3$), the cathodic adsorption of which prevents the metal dissolution and consequently affects the amount of coating formed. In the case of ODDTC, it gets converted into the acid form with simultaneous protonation of the nitrogen atom.[11] The action of ODDTC initiated through the surface adsorption of the species (resembling the onium cation) followed by its desorption as the iron tris-chelate,[11,13] ultimately increasing the amount of iron dissolved during phosphating and hence the coating weight. Being nonionic in nature, the added Span 60 exhibits non-specific adsorption affecting both the cathodic as well as the anodic partial reactions and thereby decreasing the coating weight.

4 CONCLUDING REMARKS

From the foregoing discussions it can be concluded that all three types of surface active agents - cationic, anionic and nonionic, prove to be useful in improving the performance of the phosphate coating. Moreover, they are helpful in enhancing process efficiency, a vital factor from the point of view of the finisher. Optimum concentrations of 25 mg l^{-1} of the ODA, 50 mg l^{-1} of ODDTC and 1000 mg l^{-1} Span 60 have been suggested for use in the formulated standard bath; but these concentrations may not prove to be suitable in other formulations. A knowledge of the bath chemistry in operation and the chemical nature of the surface active agent in the bath are the essential prerequisites in formulating phosphating baths using surfactants.

ACKNOWLEDGEMENT

One of us (TSNS) is grateful to the Council of Scientific and Industrial Research (CSIR), New Delhi, India, for their financial support to carry out this work.

REFERENCES

1.	W.A. Ross, Brit.Pat. 3,119, 1869.
2.	T.W. Coslett, Brit.Pat. 8,667, 1906.
3.	T.Cape, Electrocoat' 88, Cincinnati, paper 6-1, March 1988.
4.	W.Machu, <u>Korrosion u. Metallschutz</u>, 1939, <u>15</u>, 105.
5.	L.A. Leonteva, A.V.Mariorva, T.V.Tkachenko and E.A.Fomina, <u>R.Zh.Korr.i Zashch.ot Korr.</u>, 1989, 1K, 535.
6.	S.A.Balezin, F.B.Glinina, I.S. Mikhal' Chenho, <u>Zashch.Met.</u>, 1977, <u>13</u>, 30.
7.	T.S.N.Sankaranarayanan and M.Subbaiyan, <u>Surf.Coat.Technol.</u>, 1990, <u>43/44</u>, 543.
8.	T.S.N.Sankaranarayanan and M.Subbaiyan, <u>Met.Finish.</u>, 1991, <u>89</u>(9), 39.
9.	J. Rajendran, M.Sc. Thesis, University of Madras, 1991.
10.	M.M.Jones, L.T.Burka, M.E.Hunter, M.Basinger, G.Campo and A.D. Weaver, <u>J.Inorg.Nucl.Chem.</u>, 1980, <u>42</u>, 775.
11.	G.D.Thorn and R.A.Ludwig, "The Dithiocarbamates and Related Compounds", Elsevier, Amsterdam, 1962.
12.	J.I. Bregman, "Corrosion Inhibitors", Macmillan, New York, 1963.
13.	V.Srinivasan, Ph.D. Thesis, University of Madras, 1988.

Section 1.3 Wear Resistant Coatings

1.3.1
The Influence of Coatings on Wear at Elevated Temperatures

F. H. Stott[1] and D. R. G. Mitchell[2]

[1] CORROSION AND PROTECTION CENTRE, UMIST, PO BOX 88, MANCHESTER M60 1QD, UK

[2] SCHOOL OF MATERIALS AND SCIENCE AND ENGINEERING, UNIVERSITY OF NEW SOUTH WALES, KENSINGTON, NSW, AUSTRALIA

1   INTRODUCTION

Materials for applications at elevated temperatures must satisfy many criteria, often including weight, creep and fatigue resistance, strength, impact fracture toughness, weldability, formability and resistance to corrosion, erosion and wear. Improvement in some properties is only possible at the expense of others. Thus, material selection is determined by the relative importance of these criteria and, generally, a compromise is required. The material with the necessary bulk properties may not have adequate surface properties and a coating may be needed for protection.

Coatings for such applications include metals, alloys, intermetallics, ceramics and composites. However, failure of a coated system usually occurs more rapidly than expected, due to effects such as defects in the coating, inadequate throwing power of the coating process, damage to, or spallation of, the coating in service and interdiffusion between the coating and the substrate. In addition, many high-temperature alloys form in-situ coatings which can give resistance in oxidizing environments. These are established by reaction between oxygen and the alloy to form an oxide barrier layer. Although such layers may have some of the disadvantages of externally-applied coatings, they have the ability to be re-established in situ by further oxidation of the alloy if they should fail or be damaged.

In addition to chemical interactions with the environment, components at elevated temperatures can be subjected to relative motion, causing mechanical stresses, e.g. sliding wear. This can result in changes in the surface condition, particularly in the early stages when friction and wear rates can be high and 'severe wear' occurs[1]. The damage involves metal-metal contact, adhesion between contacting asperities, deformation and ploughing of the substrate. However, in environments of high oxygen activity, the development of oxide can reduce eventually metal-metal interactions and, hence, the wear rate, leading to 'mild wear'. Even at low temperatures, mild-wear conditions may be achieved, but they occur more easily and are more effective in reducing wear as the temperature is increased[2-4]. Indeed, for many alloys, under given conditions, there is a well-defined temperature above which wear-protective oxide regions (known as 'glazes') are established eventually and become effective in reducing friction and wear, and below which such regions are unable to form or be fully-effective and these parameters remain high.

Research has shown that oxide 'glazes' can be developed on most iron-, nickel- and cobalt- based high-temperature alloys, with the time required for their establishment being dependent on the conditions, particularly temperature, as discussed elsewhere[5]. For instance, they can be developed within a few seconds on Nimonic 75 at 800°C, but take 30 min to be effective at 400°C[4]. Unfortunately, wear damage can be considerable prior to their formation. In the present paper, consideration is given to the development of such in-situ oxide 'glazes' and to methods by which the metal surface may be protected in the early stages, before such 'glazes' can be established. These include pre-oxidation and external ceramic coatings.

2   EXPERIMENTAL

Tests have been undertaken in a reciprocating friction and wear apparatus, described elsewhere[1], under environmental control. The specimens were pins with hemispherical ends (5 mm radius) and discs (25 mm diameter by 3 mm thickness), usually machined from bar. In the present paper, results are presented for 321 stainless steel (18wt%Cr, 8.1%Ni, 1.1%Mn, 0.6%Ti, 0.4%Si, 0.1%Mo, balance Fe) and Jethete M152 (12%Cr, 2%Ni, 0.6%Mn, 0.3%Mn, 1.6%Mo, 0.3%V, balance Fe) under like-on-like contact. The discs were abraded while the hemispherical surfaces of the pins were ground to the required finishes (< 1 μm centre line average). Arc-evaporative physical vapour deposition was used to deposit a 2 to 3 μm thick adherent and columnar layer of titanium nitride on the surfaces of some of the 321 stainless steel specimens.

For a test, a pin was fastened with the hemispherical end in contact with a disc of the same material. Wear was achieved by a backwards-and-forwards movement of the disc specimen. For Jethete M152, the load was 15 N, the amplitude was 2.5 mm and the frequency was 8.3 Hz. The corresponding values for tests involving 321 stainless steel were 16 N, 4.8 mm and 1 Hz respectively. The specimens were surrounded by heating elements in a sealed enclosure, allowing tests at 20° to 800(±5)°C. For 321 stainless steel, the environment was de-oxygenated carbon dioxide (containing about 400 ppm water vapour and < 30 ppm oxygen) while, for Jethete M152, it was laboratory air. The coefficient of friction and contact resistance were monitored continuously by means of a strain-gauge system and a contact-resistance device respectively. Specific wear coefficients were calculated from wear volumes, measured by profilometry.

In addition, some pins and discs were pre-oxidized prior to the tests. 321 stainless steel specimens were exposed to carbon dioxide containing 1% carbon monoxide and 400 to 600 ppm water vapour at 3.1 MN m^{-2} pressure for up to 3,000 h at 700°C. Jethete M152 specimens were pre-oxidized by exposure to laboratory air for 4 h at 800°C.

3 DEVELOPMENT OF WEAR-PROTECTIVE OXIDES IN SITU

The trends in friction and wear during like-on-like sliding are relatively similar for many iron- and nickel-base alloys under various conditions. At low temperatures, average coefficients of friction and wear rates are relatively high while there is a predominance of metal-metal contact, particularly in the early stages. Friction profiles are very erratic, consistent with such contact. In the later stages, the presence of oxide on the load-bearing regions reduces, to some extent, metal-metal contact and, thus, the wear rate, giving mild-wear regimes. The friction remains relatively high while the trace continues to be irregular and erratic.

At high temperatures, oxide 'glazes' can usually develop on the load-bearing regions, giving lower friction and much reduced wear rates. Typically, metal-metal contact occurs in the early stages, unless there is sufficient oxide produced in the heating-up period prior to the wear test to prevent it. If such contact results during sliding, the initial friction and wear rates may be high. However, after a certain time, which is very reproducible for a given alloy under a given set of conditions, the coefficient of friction decreases sharply to a low and steady value at which it remains for the remainder of the test. At the same time, the wear rate decreases to a very low value, usually below detection limits, and contact-resistance measurements indicate that metal-metal contact no longer occurs. Development of smooth, compacted oxide 'glaze' regions coincides with the drop in friction and wear rate.

Figure 1 shows typical average coefficient of friction versus time plots for Jethete M152 and 321 stainless steel under the conditions specified earlier. The wear coefficients were calculated from the wear volumes measured after 10^4 s sliding, assuming a constant wear rate for a given run. As the wear rate reduces sharply to an immeasurably small value on establishment of the 'glaze', this assumption is not correct, but a low coefficient indicates formation of the 'glaze'.

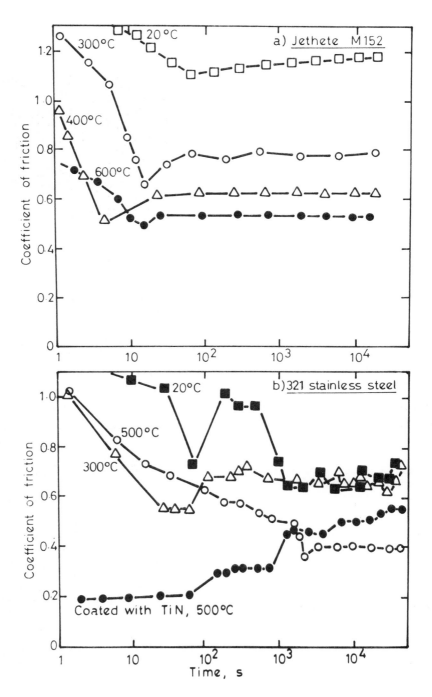

<u>Figure 1</u>   Coefficient of friction versus time plots during
          like-on-like sliding for Jethete M152   and 321
          stainless steel at various temperatures

a)   Jethete M152   b)   Jethete M152   c)   321 stainless
     for 0.5 h at   for 0.5 h at   steel, for 3 h
     300°C   400°C   at 500°C

<u>Figure 2</u>   Scanning electron micrographs of wear surfaces on
             disc specimens after like-on-like sliding

Although the trends for the two systems were similar to
each other, there were also differences.   Metal-metal
contact occurred throughout at 20°C for both;   however,
effective oxide 'glazes' developed much more rapidly on
Jethete M152 than on 321 stainless steel at 300° to 600°C.
They were established on the former within 30 s at 300°C, 10
s at 400°C and 20 s at 600°C, but, on the latter, not until
after 1,500 s at 500°C.   Although the average coefficient of
friction decreased to 0.7,   the friction trace remained
erratic and a 'glaze' was never developed on 321 stainless
steel at 300°C.   Thus, the wear coefficients for this alloy
were large, being 16 x 10^{-14} and 35 x 10^{-14} m^2 N^{-1} at 20° and
300°C respectively.   Even at 500°C, the long period prior to
formation of the 'glaze' regions resulted in a relatively
large wear coefficient, 6 x 10^{-14} m^2 N^{-1}.   For Jethete M152,
the coefficient was large at 20°C (40 x 10^{-14} m^2 N^{-1}), but
much smaller at 300°C and above (< 10^{-15} m^2 N^{-1}).

The oxide 'glazes' on many alloys have similar
morphologies.   The wear scars were covered by regions of
compacted oxide and oxide-coated metal debris, the surfaces
of which were smooth and deformed (Figs. 2(a)-(c)).
Examination using the transmission electron microscope
indicated that the surface was comprised of very fine
particles of oxide, typically 10 to 50 nm in diameter, which
had been compacted and deformed during sliding.   This
surface was often formed on large debris particles, produced
during the initial period of severe wear.   Prior to
formation of these 'glaze' regions, there was evidence for
considerable damage, typical of metal-metal contact.

The mechanisms of establishment of the 'glaze' regions
have been discussed elsewhere [5,6].   They develop on and from

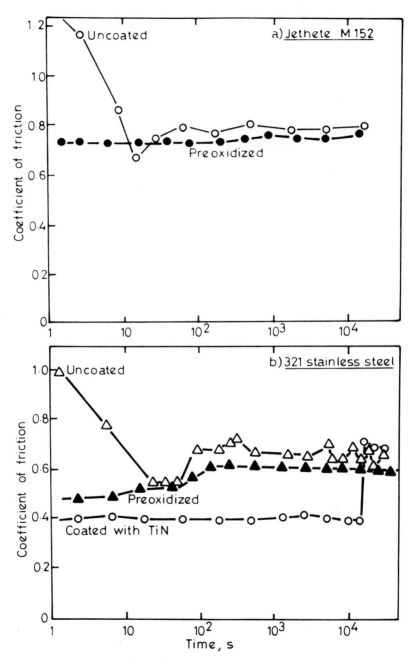

Figure 3 Coefficient of friction versus time plots during
like-on-like sliding at 300°C for as-received,
pre-oxidized and TiN-coated specimens of Jethete
M152 and 321 stainless steel

oxide and oxide-coated metal debris particles which are generated by one or more of three processes:

(i)   formation, fracture, oxidation, refracture and reoxidation of metal debris particles,
(ii)   transient oxidation of the metal surfaces, removal of such oxide on each traversal and reoxidation of the exposed alloy surface,
(iii)   growth of oxide on the metal which is not removed completely and thickens during sliding.

This debris is comminuted and compacted, with the particles at the surface being very small. The hydrostatic pressure facilitates healing of cracks in the oxide particles, resulting in flow of the asperities. Junction growth is small as the conditions for flow soon break down and failure occurs at, or near, the original asperity-asperity contacts, causing low friction and wear.

Such 'glaze' regions develop more rapidly as the rate of oxidation increases. For a given alloy, this results from an increase in temperature or frictional heating. Other important factors include alloy composition, environment, sliding speed and load. Thus, the lower alloy chromium concentration, the faster sliding speed in the tests and an environment which favours more rapid oxidation result in easier establishment of 'glaze' regions on Jethete M152 than on 321 stainless steel at a given temperature in the present tests. Oxidation rates are higher in laboratory air than in deoxygenated carbon dioxide with a low moisture content, while the rate of transient oxidation of an iron (nickel)-chromium alloy decreases as the alloy chromium concentration is increased. The faster sliding speed for the Jethete M152 tests also facilitates generation of oxide debris and, thus, formation of 'glaze' regions.

4   THE INFLUENCE OF PREOXIDATION

As 'glaze' regions are established from oxide and oxidized debris, pre-oxidation is a possible method of increasing the availability of oxide and, thus, the formation of such layers, particularly at low temperatures where they may not be developed during normal operation, or only with difficulty and after long periods. A 'glaze' surface was established almost immediately on sliding pre-oxidized Jethete M152 at 300°C and metal-metal contact was never indicated on subsequent traversals. This resulted in less wear and a lower coefficient of friction in the first few seconds of sliding compared with the corresponding as-received specimens (Fig. 3(a)).

The effects of pre-oxidation were considerable for 321 stainless steel on sliding at 300°C since 'glaze' regions were unable to develop for the as-received specimens. The pretreatment produced a thin scale consisting of an inner Cr_2O_3-rich layer and an outer iron-rich oxide, although

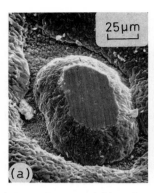 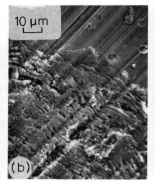 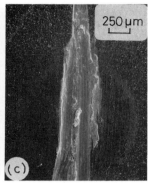

a)   Pre-oxidized   b)   TiN-coated,   c)   TiN-coated,
                            prior to   after coating
                            coating failure   failure

Figure 4   Scanning electron micrographs of wear surfaces on
            disc specimens of 321 stainless steel after like-
            on-like sliding for 3 h at 300°C

numerous thicker nodules, rich in iron, were present in
discrete locations, consistent with breakaway oxidation.
These nodules were magnetite above an iron-chromium-rich
oxide layer. On sliding at 300°C, a 'glaze' was established
almost immediately, resulting in a steady friction trace and
a low coefficient of friction (Fig. 3(b)). Metal-metal
contact was never recorded and the friction remained at a
constant value. Little wear was recorded.

Examination of the specimens revealed extensive oxide
'glaze' regions, similar to those developed on the as-
received surfaces at higher temperatures. For Jethete M152,
these consisted of deformed and compacted debris, giving
typical islands of oxide 'glaze'. They were prolific, load-
bearing and had smooth, polished surfaces, elevated above
the surrounding areas. They contained the alloying elements
in similar proportions as in the alloy. Similar features
were observed on pre-oxidized 321 stainless steel. However,
in some areas, sliding had resulted, essentially, in
polishing of the pre-formed oxide scale rather than break up
and compaction of debris from the scale (Fig. 4(a)).

5   THE INFLUENCE OF COATINGS

In addition to pre-formed oxides, externally-applied
coatings may reduce metal-metal contact in the early stages,
prior to establishment of 'glaze' regions. Tests were
carried out in which thin titanium nitride coatings were
deposited on 321 stainless steel specimens prior to sliding
at 300° and 500°C. Titanium nitride was selected because it
is a hard coating which can give wear protection at low
temperatures[7]. It is not very stable in an oxidizing
environment and may develop an oxide which may assist in
providing wear protection at these intermediate temperatures.

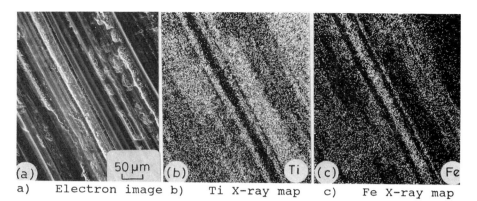

a)   Electron image b)   Ti X-ray map   c)   Fe X-ray map

<u>Figure 5</u>   Scanning electron micrographs of wear surface on disc specimen of TiN-coated 321 stainless steel after like-on-like sliding for 12 h at 500°C

During sliding at 300°C, the friction versus time profiles were different from those for the uncoated specimens (Fig. 3(b)). The trace was very steady and the coefficient of friction was approximately 0.4 for about 2 h. However, subsequently, the value increased rapidly to about 0.7. At the same time, the trace became very irregular for the rest of the test. The experiment was repeated and similar traces were recorded, with the transition to the high value always occurring after between 100 and 200 min. Prior to the transition, the only wear damage was polishing of the coating surface. However, the coating had been progressively thinned with time in this period until it started to crack and break up (Fig. 4(b)). Eventually, underlying substrate was exposed and metal-metal contact ensued (Fig 4(c)). This led to rapid failure of the remaining coating and the friction trace became similar to that for the uncoated specimens. After the transition, the wear rate became much higher, with typical deformation and ploughing damage. Hence, although the coating had given protection in the early stages, it did not facilitate development of 'glaze' regions.

Similar tests were carried out at 500°C, where uncoated specimens had suffered significant damage prior to development of the 'glaze' surfaces. For this system, the pre-'glaze' period was about 1,500 s. The presence of the titanium nitride coating resulted in a smooth friction trace in the early stages and a low coefficient of friction (Fig. 1(b)). Thereafter, it increased with time, to about 0.5 after 900 s and 0.55 after 10 h. Moreover, the trace remained steady throughout. Little wear was detected after the test, but there were deep grooves in the coating which penetrated to the substrate. However, there was no failure of the coating and oxidation of the exposed substrate prevented metal-metal contact (Fig. 5(a)). The resulting surface was essentially an oxide 'glaze' consisting of areas of oxidized coating, probably TiO_2, and areas of oxidized iron-rich substrate where the coating had been penetrated (Figs. 5(b) and (c)). Hence, at this temperature, the coating had been effective in giving protection in the early stages while wear-protective oxide, formed in situ, continued to give protection in the later stages.

6   CONCLUSIONS

1. Metal surfaces can be protected against wear damage during like-on-like sliding by the in-situ formation of wear-protective oxide 'glazes' following reaction with the environment, particularly at high temperatures.
2. Pre-oxidation of the surfaces can reduce metal-metal contact in the early stages and provide an additional supply of oxide debris, leading to more rapid establishment of the 'glaze' surfaces.
3. Externally-applied thin ceramic coatings can also prevent metal-metal contact in the early stages. However, such coatings are progressively thinned by the sliding action until failure occurs. Subsequent wear damage is determined by the ability of the exposed metal to develop the oxide 'glaze' surfaces.

ACKNOWLEDGEMENTS

The authors thank the SERC for a Research Studentship (to DRGM) and Nuclear Electric, particularly Dr. T.C. Chivers and Dr. J. Skinner, for advice and support under the CASE scheme. They are grateful to Dr. J. Glascott who obtained the Jethete M152 data as part of his PhD programme.

REFERENCES

1. D.S. Lin, F.H. Stott, G.C. Wood, K.W. Wright and J.H. Allen, Wear, 1973, 24, 261.
2. D.H. Buckley and R.L. Johnson, ASLE Trans., 1960, 3, 93.
3. T.F.J. Quinn, ASLE Trans., 1978, 21, 78.
4. F.H. Stott and G.C. Wood, Tribology, 1978, 11, 211.
5. F.H. Stott, J. Glascott and G.C. Wood, Corrosion-Erosion-Wear of Materials at Elevated Temperatures, ed. A.V. Levy, 1987, 263, National Association of Corrosion Engineers, Houston, Texas.
6. J. Glascott, F.H. Stott and G.C. Wood, Wear, 1984, 97, 145.
7. E.S. Hamel, Materials Engineering, 1986, 103, 8.

1.3.2
Influence of Two-phase Lubricants on Slight Movement Wear Characteristics

S. Baoyu,[1] L. Xinyuan,[1] Z. Wenjie,[1] Q. Yulin,[1] J. Zhishan,[2] and L. Hong[2]

[1] DEPARTMENT OF MECHANICAL ENGINEERING, HARBIN INSTITUTE OF TECHNOLOGY, PEOPLE'S REPUBLIC OF CHINA

[2] LANZHOU INSTITUTE OF CHEMISTRY AND PHYSICS, LANZHOU, PEOPLE'S REPUBLIC OF CHINA

1 INTRODUCTION

The two-phase lubricant is a new development in which a solid powder is added to a liquid lubricant. Advantages characteristic of both liquid lubricant and solid lubricant enable the achievement of better tribological properties, such as antifriction, antiwear and contact fatigue life. This paper describes the measurement of slight movement wear in the two-phase lubricant containing various levels of graphite or carbon or PTFE powder.

2 EXPERIMENTAL DETAILS

Test Oil Samples

No. 20 machinery oil was used as the base oil. The diameters of the graphite, carbon and PTFE powders were all below 1-2 µm. Graphite, carbon and PTFE powder have characteristic layered, irregular and high polymer structures respectively. A dispersant was used to suspend the solid powder in the oil - see Table 1.

Test Rig and Test Pieces

The SRV slight movement wear test rig was used in this work. The configuration of the test piece - comprising a GCr15 steel ball of diameter 10mm and hardness HRC 60-63, and No. 45 steel flat plate of hardness HB 250-260 - is shown in Figure 1.

Table 1 Test oil samples

Sample No.	Content of solid powder in No. 20 machinery oil (Weight ratio)
A1	graphite 0.5%
A2	graphite 1.0%
B1	carbon 1.0%
B2	carbon 2.0%
C1	PTFE 1.0%
C2	PTFE 5.0%

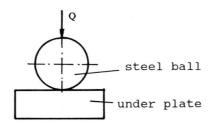

Figure 1 Schematic of test piece

Test Conditions and Procedure

The load, horizontal amplitude and vibration frequency were all fixed at 100N, 1000μm and 50 Hz.   All
experiments were carried out at 20^{0}C and lasted 25 minutes.

Before testing, the specimens were cleaned in acetone
and dried, installed in the test rig and had test oil
applied to the plate.

3 RESULTS

Friction Coefficient

Figure 2 records the variation of friction
coefficient with time for test lubricants A1, B1, and C1.
The friction coefficient of the lubricant containing
graphite (A1) is small and changes little with time.   That

of the lubricant containing carbon powder (B1) is also
small, but tends to decrease with time. In contrast
lubricant C1, containing PTFE, yields a large value of f
which tends to increase with time.

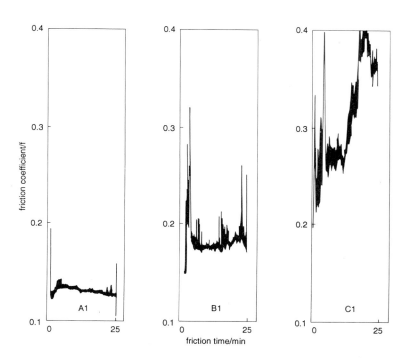

Figure 2 Changes in coefficient of friction over test
duration for lubricants containing graphite
(0.5 wt%) A1, carbon (1.0 wt%) B1, and PTFE
(1.0 wt%) C1

Wear Quantity

After testing, the quantity of test plate wear
was estimated using a surface profilometer. Figure 3
shows the range of wear volumes for the test conditions,
with the ranking extending from small for graphite
additions (A1 and A2) through to relatively high for
carbon (B1 and B2) and PTFE (C1 and C2) additions.
Variations in wear volumes with levels of additives are
clearly evident.

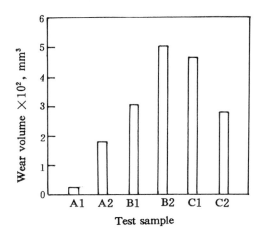

<u>Figure 3</u> Wear volumes as a function of levels and types of
        additives

Surface Topography

Post-test examination by optical microscopy of the
under test pieces yielded micrographs similar to those
shown in Figure 4 - here the thickness of the surface
scratch for the Al test piece is small, that for Bl is
broad and that for Cl lies in-between.

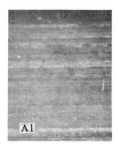 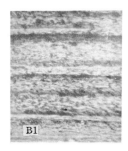 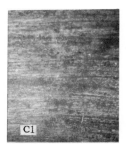

<u>Figure 4</u> Optical photomicrographs of wear surfaces

4 DISCUSSION

The addition of graphite powder, due to its layered
structure, led to small coefficients of friction, the
values of which changed little with changing levels of
graphite. In contrast supplements of carbon powder,

because of its irregular crystalline structure, produces
large values of f. The friction coefficients for PTFE
additions are even larger and vary considerably with time
and levels of addition - here under large contact
pressures the PTFE powder is easily deformed and metal-
metal contact is extensive.

Wear characteristics followed analogous patterns to
the friction tests.

5 CONCLUSION

Under the specified test conditions certain deductions may
be made:-
1. Very low values of friction coefficient were recorded
   if contacting metal surfaces were separated by
   graphite which is characterized by a layered
   structure.
2. Contact surfaces separated by carbon powder showed
   considerable abrasion-related wear.
3. Extensive metal-metal contact and high levels of
   friction and wear occurred when PTFE powders were
   deformed between touching ball and plate. Values of
   friction and wear decreased as the level of PTFE
   addition increased.

REFERENCES

1.   S. Baoyu et al., Proc. Japan Int. Trib. Conf.,
     Nagoya, 1990, p 11.
2.   S. Baoyu et al., Proc. Germany Int. Trib. Conf.,
     1992, p 1.
3.   M. Matsunaga et al., 'Handbook of Solid Lubrication',
     Chinese translation, published by Mechanical Industry
     Press, 1986.
4.   Q. Yulin, 'Friction and Wear' (In Chinese), The
     Academic Education Press, 1986.

1.3.3
Dry Friction and Wear Behaviour of a Complex Treatment Layer

Y. Wang[1] and Z. D. Chen[2]

[1] MATERIALS SCIENCE AND ENGINEERING DEPARTMENT, BEIJING UNIVERSITY OF AERONAUTICS AND ASTRONAUTICS, BEIJING 100083, PEOPLE'S REPUBLIC OF CHINA

[2] MATERIALS SCIENCE AND ENGINEERING DEPARTMENT, TIANJIN UNIVERSITY, TIANJIN 300072, PEOPLE'S REPUBLIC OF CHINA

```
1 INTRODUCTION
```

It is well known that friction and wear processes occur primarily at the surface of solids so the surface characteristics play an important rôle. Thermochemical treatment is an effective technical measure to increase the wear resistance of steels because it can improve the chemical compositions, structures and properties of steel surfaces. The common methods of surface hardening treatments, carburizing, nitriding and carbonitriding are of obvious effect to improve the strength and toughness of steels, whilst different types of liquid lubricants are generally used in tribological systems in order to reduce or minimize wear and to prevent surface scuffing and seizure between mating surfaces. Scoring, galling, scuffing and seizing are still the main failure modes of numerous mechanical components which undergo surface hardening treatments.

Because the Sulf-BT process can decrease the coefficient of friction and increase the wear resistance of steel, the method has received much attention. This method has been reported to solve difficult lubrication problems and some actual applications have been described[1-4]. The method can be used for various irons and steels because of its many advantages i.e. the process is characterized as low-temperature, rapid, pollution-free and with no dangers of hydrogen embrittlement etc. In order to increase effectively the wear resistance of mechanical components, we put forward a surface-strengthened and lubricated complex thermochemical treatment. Its aim is to establish an ideal tribo-surface layer which has an optimum combination of a high strength matrix and low shear interface.

Friction and wear are very complicated problems involving a series of influencing factors. From the aspect of engineering applications, the combination or compatibility of two contacting components is probably one of the most important parameters[5]. So, the structures and properties of steel surface layers after different surface treatments, their effects on friction and wear and their dynamic behaviour in friction and wear processes should be studied systematically. Since this surface treatment is still at the development stage, suitable engineering application conditions are still not specified; a charac-terization of the wear behaviour of surface layer is necessary[3].

The surface structures and the friction and wear behaviour of steel $20Cr_2Ni_4A$ after complex thermochemical treatment combining carburizing and low-temperature electrolytic sulphurizing, i.e. the Sulf-BT process, were studied by comparative testing. The purpose of the present study is to approach the effect of such complex treatment on friction and wear processes of steel and to clarify the friction and wear mechanisms.

2 EXPERIMENTAL PROCEDURE

The material used was steel $20Cr_2Ni_4A$. The composition of the material is shown in Table 1. All pins were of dimensions 10x4x20mm and rings of 32 mm diameter and 10 mm thickness and underwent the following pretreatment:

(a) Carburizing at 900 ± 100^0C for 17 hrs, case-depth is about 1.3-1.6mm; (b) tempering at 650^0C for 5.5 hrs; (c) heating at 830^0C in a salt bath for 8 min., with oil quenching and then tempering at 150^0C for 3 hrs - a hardness of HRC62-64 was obtained; (d) after grinding, a roughness of Ra=0.6μm was obtained. Half of the carburized specimens underwent electrolytic sulphurizing after degreasing and pickling. Electrolytic sulphurizing was carried out in a molten salt bath with bath composition of 75%KSCN + 25%NaSCN. The bath temperature was 190 ± 10^0C. Duration of treatments was 20 min. Treated specimens were set as anodes; the cathode was stainless steel. The

density of the anodic current was 2.5A/dm^2. After
sulphurizing and rinsing the specimens in running water to
dissolve the frozen salt crust, they were then put into
engine oil.

The thickness of the sulphurizing layer was measured
by physical and metallographical methods. The replica
structure of the sulphurizing layer was analysed with a
model DXA-10 transmission electron microscope. The
distribution of sulphur in the surface layer of
sulphurized specimens was examined with a model XW-01
electron probe microanalyser. All friction and wear tests
were performed on a pin-on-ring tester without lubrication
at room temperature. The test parameters were as follows:

Normal load: 50-250N;
Sliding speeds: 1, 1.5, 2m/s;
Sliding distance: 600m.

The frictional coefficient was recorded. Weight loss
was measured by a precision balance to 0.1 mg. Every test
parameter was repeated with four sets of pin and ring
specimens. Worn surface morphology and wear particles
were observed by using type S-550 and type JSM-35C
scanning electron microscopes.

3 RESULTS AND DISCUSSION

Structure of the Complex Treatment Layer

A thin (several micron thick) FeS film can be formed
on the surface of carburized specimens after electrolytic
sulphurizing at 190 ± 10^{0}C for 10-30 min. Figure 1 and
Figure 2 show the SEM and TEM morphology of the
sulphurised layer. It can be seen that the FeS film
possesses a cloud layer-like, or scale-like morphology.

Figure 3 shows the absorption electron image and the
SK$_\alpha$ concentration of the sulphide layer.

Frictional Coefficient

According to the adhesion theory of metals with
contaminant films, the coefficient of friction can be
written as[6]:

Table 1   Composition of material used

Element (%)	C	Si	Mn	Cr	Ni	Fe
Pin material	0.17	0.31	0.30	1.61	3.65	bal.
Ring material	0.17	0.30	0.49	1.63	3.71	bal.

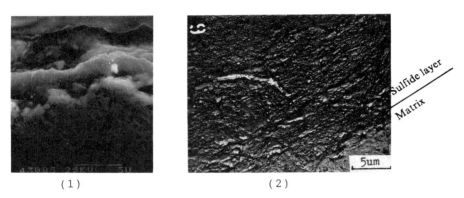

(1)   (2)

Figure 1   SEM morphology of the sulfurized layer
Figure 2   TEM morphology of the sulfurized layer

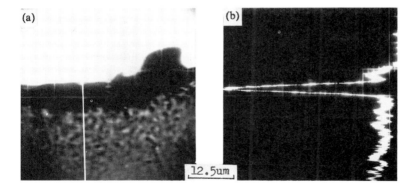

Figure 3 Absorption electron image (a) and the SK_α
concentration spectrum (b) of sulphide

$$\mu = \frac{\text{critical shear stress of the interface}}{\text{yield pressure of the bulk metal}} \qquad (1)$$

Based on this equation, it is reasonable to strengthen the matrix of steel through carburizing and make a surface layer with lower shear strength through sulphurizing treatment. Thus an ideal complex surface layer can be formed. Because the sulphurized layer (FeS) is a non-metallic inorganic substance with a close-packed hexagonal lattice, and has a lamellar and porous oil-retaining structure which slides very easily, it can exhibit excellent friction-reducing properties whether in a lubricated or unlubricated condition[7]. Many studies have proved that the sulphurized layer is an excellent solid lubricant[3,4]. Figure 4 shows the variations of the frictional coefficient of carburized and carburized plus sulphurized specimens with sliding distance under a speed of 2 m/s. It can be seen that, sulphurizing after carburizing effectively decreases the frictional coefficient (0.35-0.65) of carburized specimens to a value of 0.05-0.2 (see the dashed line in Figure 4). It can be also seen that the sulphurized layer can effectively promote the running-in process of the rubbing pairs. Therefore, their contact conditions can be rapidly improved and enter the steady stage of wear.

Wear and Wear Rate

The wear resistant properties can be reflected in the weight loss or wear rate of the worn pin specimens. Figure 5 displays the weight loss of carburized and carburized-sulphurized pin specimens for various test conditions. From this it can be seen that the carburized pins show greatly increased weight loss with sliding speed, but the carburized-sulphurized specimens have very slight weight loss almost independent of the test conditions.

Figure 6 shows the weight change of ring specimens before and after sliding wear. It is interesting that the weight of carburized rings did not decrease but increased during sliding. This phenomenon is the result of metal transfer during sliding and is an obvious characteristic of adhesive wear. Figure 7 records the wear rate of

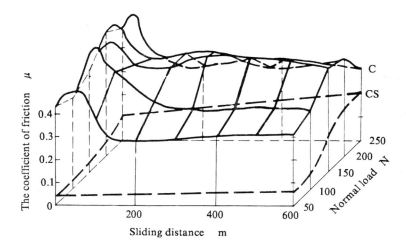

Figure 4 Frictional coefficient of carburized (C) and carburized— sulfurized (CS)
specimens with sliding distance under the condition of 2m / s speed

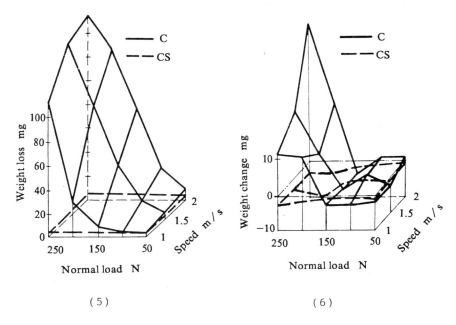

(5)   (6)

Figure 5 Weight loss of different pin specimens
Figure 6 Weight change of different ring specimens

carburized and carburized-sulphurized pin specimens at various testing conditions. It indicates the superiority of the complex treatment over single carburizing. It also reveals that a transition from mild wear to severe wear occurs with an increase in load and speed beyond a certain value. It has been suggested that if the wear rate, Ws, is less than $10^{-8}mm^3$ / (N.mm), then wear can be considered as mild wear. However, if the wear rate, Ws is greater than $10^{-8}mm^3$ / (N.mm), then wear can be considered as severe wear[8]. When the wear rate > $10^{-8}mm^3$ / (N.mm) with increasing load or speed, the wear mechanism of carburized specimens is severe (adhesion dominated) wear. In this case, the surface flash temperature and the surface mean temperature are increased so that the strength of the material in the surface layer will be reduced. As a result, plastic deformation and metal transfer will easily occur during sliding. Thus, the weights of carburized pin specimens decrease, while that of carburized rings increase. Under the same conditions, the weight of carburized-sulphurized specimens shows almost no changes and the wear mechanism is still mild wear.

Worn Surfaces and Wear Particles

Worn surface morphology will be changed due to the transition from mild wear to severe wear during sliding. As worn surfaces are roughened, grooves are deepened, adhesive junctions are formed, fatigue cracks appear and various typical products (e.g. scale-like features characteristic of severe adhesive wear) can be found (as shown in Figure 8). Because the sulphide layer has excellent friction-reducing and wear resistant properties, and can separate the wearing surfaces from direct metallic contact, the worn surfaces of all carburized-sulphurized specimens are comparatively smooth (see Figure 9).

Figure 10(a) shows some wear particles of carburized specimens which are from sliding wear under 50N load and 2 m/s speed. In this case, most of the wear particles are thin flakes, and are considered a product of oxidational wear. However, the three-dimensional size of wear parti-cles increases with increasing load and speed. Figure 10(b) shows wear particles of carburized specimens produced from sliding under 150N load and 2 m/s speed. Thick and large flakes were found which indicate onset of severe wear. In general, it should be emphasized that

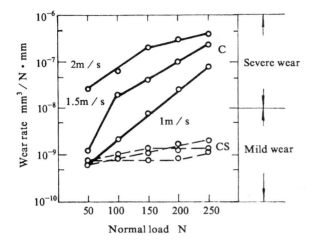

Figure 7   Wear rate of different pin specimens

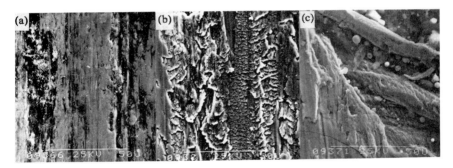

Figure 8   Worn surfaces of carburized ring specimens.
    (a)50N; (b) 200N; (c) 250N

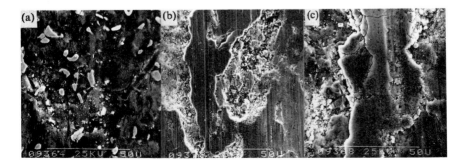

Figure 9   Worn surfaces of carburized–sulfurized ring specimens.
    (a) 50N; (b) 200N; (c) 250N

steel-to-steel wear is not under the action of one single
wear mechanism. Sometimes adhesive wear is accompanied by
abrasive wear. At that time wear particles and adhesive
transferred metals that were work hardened or quench
hardened often act as abrasives to cut the surface of
materials. Wear particles then take the form of spirals,
loops and bent wires. Figure 10(c) is a micrograph of
wear particles produced from sliding under the conditions
of 250N load and 2 m/s speed. A typical curling product
can be noted.

As compared with carburized specimens, a small amount
of wear particles is produced during sliding for carburized plus sulphurized specimens. Some of the wear particles are fine flake-like sheets which can be considered as
from metal asperities, and others are very tiny black wear
particles which can be considered as from brittle and weak
sulphide layers (as shown in Figure 11). These phenomena
indicated the characteristics of mild wear.

Observations of Worn Surface Layers

Figure 12 shows the optical micrographs of metallographic sections of the worn surface layers of
differently treated pin specimens. It can be seen from
the photographs that under the conditions of 50N load and
2 m/s speed, there is obvious plastic deformation and a
white etched surface layer in the worn surface layer of
the carburized pin specimen. However, there is no obvious
plastic deformation nor white-etched surface layer to be
seen in the worn surface layer of the carburized-
sulphurized specimen, see Figures 12(a) and (b). At
higher load, the thickness of the plastically deformed
layer in the worn surface layers of carburized specimens
is larger than that of carburized-sulphurized specimens,
as shown in Figures 12(c) and (d).

Microhardness Distribution in Worn Surface Layers

Figure 13 shows the microhardness distribution curves
of the worn surface layers of both the carburized and
carburized-sulphurized pin specimens at a speed of 2 m/s.
It can be seen that the microhardness in worn surface
layers of carburized pins changes remarkably with
increasing load. Below 150N load, the microhardness in
worn surface layers decreases due to friction softening.

Figure 10 Wear particles of carburized specimens
(a) 50N; (b) 150N; (c) 250N

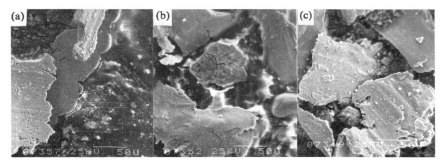

Figure 11 Wear particles of carburized—sulfurized specimens
(a) 50N; (b) 150N; (c) 250N

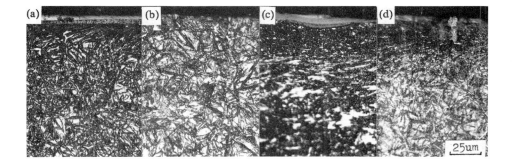

Figure 12 Worn surface layers of different pin specimens.
(a) carburized pin, 50N; (b) carburized—sulfurized pin , 50N;
(c) carburized pin, 250N;(d) carburized—sulfurized pin, 250N

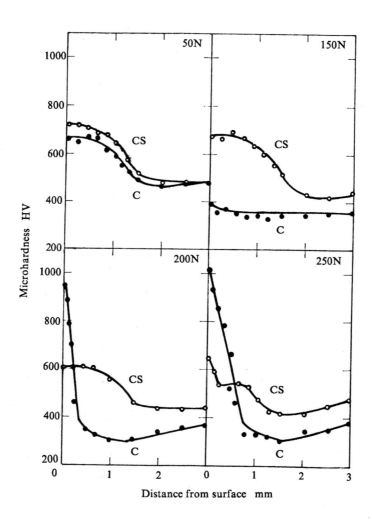

Figure 13   Microhardness in worn surface layers

When the testing load is above 200N, the microhardness in most worn surface layers increases due to the white-etching surface layers being developed. However, because the sulphide layer can prevent changes in e.g. the temperature, structure and property in surface layers, during sliding, the microhardness in worn surface layers of carburized-sulphurized specimens is not obviously decreased. That is to say, the load-carrying capacity is improved by such a complex treatment.

4 CONCLUSIONS

1. A surface-strengthened and lubricated complex thermochemical treatment was put forward. After such treatment, an ideal tribo-surface layer which has an optimum combination of a high strength matrix and low shear stress interface can be established.

2. A sulphide layer can be formed by low temperature electrolytic sulphurizing at $190 \pm 10^0 C$ for about 20 min. The sulphide layer, deposited as lamellae on the carburized layer, has an hexagonal structure, reduced hardness and low coefficient of friction.

3. The sulphide layer not only separates the wearing surface from direct metallic contact as a solid lubricating layer but also checks dynamic changes in surface layers during sliding. As a result, the severe wear of carburized specimens can be replaced by the mild wear of carburized-sulphurized samples. The load-carrying-capacity is also improved by such a complex treatment.

REFERENCES

1. French Patent: 2050754.
2. N. Kuwayama, J. of the Japan Society for Heat Treatment, 1972, 12,(4), 250.
3. V. Veronesit, Tribology International, 1985, 18,(4), 203.
4. J. C. Gregory, Tribology International, 1987, 11,(2), 105.
5. P. L. Ko, Tribology International, 1987, 20,(2), 66.
6. J. Halling, 'Principles of Tribology,' The Macmillan Press Ltd, London, 1975.

7. E. Cosmacini and V. Veronesit, <u>Wear</u>, 1987, <u>73</u>,(1), 1.
8. E. Takeuchi, <u>Machine and Tools</u>, 1979, <u>23</u>,(5), 5.

1.3.4
The Microstructure and Wear Resistance of Laser-remelted $M_{80}S_{20}$ Alloy Coatings with or without Rare Earth Elements

Y. Wang[1] and J. J. Liu[2]

[1] MATERIALS SCIENCE AND ENGINEERING DEPARTMENT, BEIJING UNIVERSITY OF AERONAUTICS AND ASTRONAUTICS, BEIJING 100083, PEOPLE'S REPUBLIC OF CHINA

[2] TRIBOLOGY RESEARCH INSTITUTE, TSINGHUA UNIVERSITY, BEIJING 100084, PEOPLE'S REPUBLIC OF CHINA

1 INTRODUCTION

Laser modification as one of the important surface techniques has developed rapidly in the last decade[1-3]. Laser-remelting of a thermal sprayed alloy coating can improve its compactness, bonding strength with substrate, wear and corrosion resistance. Therefore, it is increasingly becoming one of the most popular laser modification technologies[4,5].

The favourable rôle of rare earth elements in steels and their application in chemical heat treatment of steel have often been reported[6-8]. However, there is still a lack of publications about the application of rare earth elements in laser modification technology.

This paper, aiming to provide the experimental basis for practical applications, studied the effect of rare earth oxide CeO_2 on microstructure and wear resistance of thermal sprayed + laser-remelted $M_{80}S_{20}$ self-fluxing alloy coatings.

2 MATERIALS AND EXPERIMENTAL METHODS

The main compositions of the iron-base amorphous self-fluxing alloy powders of 150 mesh used in this work are, in at%, (65-70)Fe, (3-5)Cr, (2-4)Ni, (2-4)W, (1-2)Mo, (10-14)B, (4-7)Si and (2-3)C. Because the atomic ratio of metal-metalloid is about 80:20, so this alloy type is abbreviated to $M_{80}S_{20}$[9].

The coating material is 1020 steel, 880^0C water quenched and 180^0C tempered to a low carbon martensite structure with HRC 35-45 hardness.

After cleaning, shot blasting and preheating, a thin layer of about 0.1-0.15mm thickness Ni-Al alloy was sprayed on the specimen using an oxygen-acetylene torch for better bonding of coating with the substrate. Then the $M_{80}S_{20}$ and $M_{80}S_{20}$ + 8%CeO_2 alloy coatings were sprayed to about 0.6-0.8mm thickness.

The single pass remelting process was conducted using a CO_2 laser of 5 kW power. The parameters selected in this research were: 2.8 kW power, 1x4mm beam spot and different traverse speeds.

The distribution of microhardness along the depth of coating was measured using a Vickers microhardness indenter and a load of 100g. The microstructure and composition of the coating were analysed by means of a scanning electron microscope and electron microprobe.

The wear tests were performed on an Amsler testing machine using the block and ring specimens. The dimensions of the block specimen were 8x8x18mm. The counterpart ring specimen was made of 52100 bearing steel, 840^0C oil quenched and 180^0C tempered with HRC 60 hardness. The wear tests were carried out under 300N load and 0.8 m/s sliding speed without lubrication. The weight loss of the block specimen was measured and converted to wear volume and wear rate. The mean value of three measurements was taken as the experimental result.

3 RESULTS AND ANALYSES

Surface Characteristics of Coatings

Due to the large surface tension, molten $M_{80}S_{20}$ alloy showed unfavourable wettability on the steel surface. Only appropriate parameters can guarantee the surface quality of such a coating. In this research, the 10mm/s traverse speed of the laser beam was desirable for achieving a better coating surface. If the traverse speed was up to 30mm/s, it tended to form molten droplets and an uneven surface. However, the rare earth element can increase the fluidity of the molten alloy and reduce its surface tension, thus a more continuous and smooth coating surface of $M_{80}S_{20}$ + 8%CeO_2 alloy can be obtained.

Metallographic Section of the Molten Pool

Usually, the dimension of the molten pool is reduced along with an increase of traverse speed due to the decrease of energy provided by the laser beam to the molten pool.

The addition of CeO_2 can enlarge the width of the molten pool and reduce its depth because of the improved fluidity and wettability of the molten alloy. Such a pool is desirable for obtaining a more even and smooth coating surface.

Microstructure of the Laser-remelted Zone

Figure 1 shows micrographs of the laser-remelted zone for 10mm/s and 30mm/s traverse speeds. The following can be deduced:

1. Along with the increase of traverse speed both $M_{80}S_{20}$ and $M_{80}S_{20}$ + 8%CeO_2 coatings show an obvious refined microstructure and reduction of spacing in the secondary dendrite structure. The addition of CeO_2 makes the refining effect more significant. Figure 2 shows the relationship between the spacing of secondary dendrite structure and traverse speed.
2. The addition of CeO_2 can improve the morphology and distribution of eutectics and compounds. The compounds are obviously rich at grain and dendrite boundaries. The gross dendrite structure has been replaced by fine equiaxed eutectics at the higher traverse speed. The refined dendrite boundaries form a more compact network structure.
3. The microvoids in the melting zone are obviously reduced due to the effect of the rare earth element.
4. The dimensions of martensite are also decreased.

Distribution of Compositions in the Laser-remelted Zone

The result of electron microprobe analysis for various compositions in the laser-remelted zone shows that their contents are higher than in the matrix. The amount of Ce is very small in the $M_{80}S_{20}$ + 8%CeO_2 coating; not more than 0.5% can be detected.

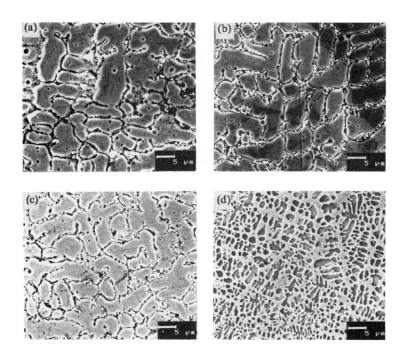

<u>Figure 1</u> SEM morphologies of microstructures of $M_{80}S_{20}$
         and $M_{80}S_{20}+8\%CeO_2$ alloy after laser-remelting:
         (a) $M_{80}S_{20}$, 10mm/s; (b) $M_{80}S_{20}+8\%CeO_2$, 10mm/s;
         (c) $M_{80}S_{20}$, 30mm/s; (d) $M_{80}S_{20}+8\%CeO_2$, 30mm/s

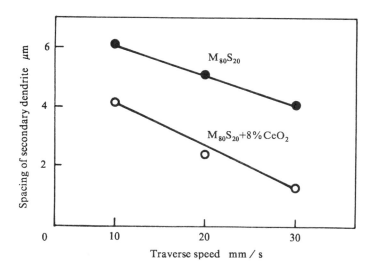

<u>Figure 2</u> Relation between spacing of secondary dendrite and
laser traverse speed

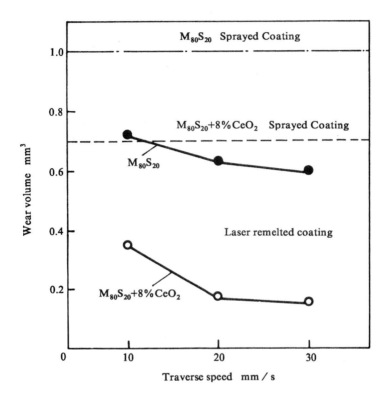

<u>Figure 3</u> Wear volume of laser-remelted coatings

Microhardness in the Laser-remelted Zone

The addition of CeO_2 can increase significantly the microhardness of a coating due to its effect of refining grains. The microhardness for 20 and 30mm/s traverse speeds already approaches that of the original amorphous structure. HV_{100} is about 650-900.

Wear Resistance of Laser-remelted Coatings

Figure 3 shows the wear volume of laser-remelted coatings for different traverse speeds. As a comparison, the wear volume of a sprayed coating is also illustrated. It can be seen that the addition of CeO_2 can increase the wear resistance quite significantly. The wear resistance of $M_{80}S_{20}$ + 8%CeO_2 coating is 2-3.5 times that of $M_{80}S_{20}$ coating after laser-remelting. If compared with the conventionally treated 52100 steel, the wear resistance of laser-remelted $M_{80}S_{20}$ + 8%CeO_2 alloy can be 12-30 times that of 52100 steel. The wear tracks show that the laser-remelted $M_{80}S_{20}$ + 8%CeO_2 coating experienced only mild wear.

4 CONCLUSION

Based on the comparison and analyses of structure and properties of laser remelted $M_{80}S_{20}$ and $M_{80}S_{20}$ + 8%CeO_2 coatings it is obvious that the addition of the rare earth element can indeed improve the surface quality, microstructure, microhardness and wear resistance of the laser modified layer. Such a result is of great scientific and practical significance.

REFERENCES

1. C.W. Draper and J.M. Poate, International Metal Reviews, 1985, 30,(2), 85.
2. T. Bell, Surface Engineering, 1987, 3,(4), 271.
3. A. Galerie, M. Pons and M. Caillet, Materials Science and Technology, 1989, 5,(8), 806.
4. S. Dallaire and P. Cielo, Metallurgical Transactions B, 1982, 13,(B9), 479.
5. M. Boas and M. Bamberger, Wear, 1988, 126, 197.
6. P.E. Waudby, International Metal Review, 1978, 23,(2), 74.
7. Z.S. Yu, Y.Y. Chu et al, 'Rare Earth Elements in Steels,' China Metallurgy Press, 1982.

8. J.S. Yu and Z.S. Yu, 'The Application of Rare Earth Elements in Iron and Steels,' China Metallurgy Press, 1987.
9. J.Y. Zhang, C. Liu and B.R. Ai, in 'Rapidly Solidified Materials,' Edited by P.W. Lee and R.S. Carbonara, ASM, Ohio, 1986, p 179.

Section 1.4 Organic Coatings

1.4.1
Critical Assessment of Electrical Impedance Spectroscopy for the Evaluation of Corrosion Protective Properties of Organic Coatings

W. Funke

UNIVERSITY OF STUTTGART AND FORSCHUNGSINSTITUT FÜR
PIGMENTE UND LACKE E.V., STUTTGART, GERMANY

1 INTRODUCTION

In the atmospheric corrosion of metals, electrochemical processes play a dominant rôle. Accordingly electrochemical testing methods have been widely used for a long time to study the corrosion behaviour of metals[1]. The evaluation of corrosion protective properties of organic coatings by electrochemical methods also has a history of more than 50 years.[2] In the subsequent time much effort has been devoted to proving that electrochemical data are useful to predict the practical performance of corrosion protective coatings.[3-6]

More recently impedance measurements with variable frequencies - electrical impedance spectroscopy (EIS) - became the method of preference for scientific studies to test and evaluate the corrosion protection by organic coatings as is illustrated by a large number of publications.[5,7-16]

However, quite a number of comments and papers are known[17-24] which warn against an uncritical use of electro-chemical data to predict corrosion protective properties of organic coatings. Therefore paint technicians in general still hesitate to use electro-chemical data and also EIS as a decisive criteria for the practical usefulness of organic coatings. This situation is also reflected by the fact that such tests still have not been standardized as practical tests to be recommended for the classification of organic protective coatings.

2 PROBLEMS IN APPLYING EIS FOR CORROSION TESTS

There are a number of problems in using EIS data for evaluating the protective quality of organic coatings to which not enough attention has been paid.

It is frequently claimed that EIS data compare well with usual standardized corrosion tests, such as the salt spray test.[25] However, these tests only allow a rough classification and their results do not always agree with practical performance. As a consequence of this, alternating exposure tests for car coatings[26-31] have been developed which better imitate practical exposure conditions. More recently it could be demonstrated, also by impedance measurements,[14] that cyclic exposure results in a fast degradation of the coating system as compared with static immersion in aqueous solutions of sodium chloride.

Contrary to basic protective properties of organic coatings, such as permeability and wet adhesion, electrical data of EIS and other electrochemical methods have a classificatory character and depend in a more complex way on the basic properties. An essential parameter in EIS is mobility and diffusion of ions in organic coatings under an external electrical potential. Apart from the presence of pores and similar coating defects it is still controversial whether ions may penetrate intact organic coatings which have a high ohmic resistance and protect well against corrosion if no external electrical potential is applied.

Osmotic blistering depends on the semipermeability of organic coating films. This phenomenon would not be possible if ions could penetrate the film. Probably inherent electrical charges of coating films contribute to this barrier effect for ions.

In cathodic protection it is imperative that coating systems completely isolate the metal surface. Otherwise cathodic delamination would occur. Therefore, the voltage amplitude chosen in EIS is very low (10-50mV) as compared with cathodic protection of steel (U_H = -530mV) and it is assumed that this external potential does not influence the normal corrosion process, not even in the low frequency range, where conditions approach those of DC experiments.

In EIS the potential gradient is usually applied across the coating film. In underrusting of organic coatings, however, the electrical potential exists between anodic and cathodic areas along the metal surface, i.e. parallel to the lateral extension of the film. Considering the high ohmic resistance of protective coating systems (10^8-$10^{10}\Omega$), it is improbable that the ion transport taking place in the corrosion elements at the metal surface goes across the film instead of along the coating/metal interface. This means that the electric circuit in EIS is different from that of the normal corrosion process.

As in all electrochemical measurements in EIS, electrolytes are needed. Usual electrolytes are salts containing anions such as sulphate and chloride. Both of them are well known as effective corrosion stimulants. This is not disadvantageous if the practical exposure conditions also involve such anions e.g. sea water, deicing salts on roads in winter or contaminated air (acid rain). As a remedy for other exposure conditions it was proposed to reduce the electrolyte concentration to a minimum, to use less aggressive electrolytes, such as potassium nitrate, or to expose the coating system to the electrolyte only during the short time of EIS measurements.[15]

The corrosion of bare as well as protected steel is a complex and dynamic process. Depending on the exposure conditions a number of different corrosion products may be formed with varying stability.[32-34] Some of these products, which appear intermediately, are iron oxide complexes, which may form colloidal membranes. These membranes have electrical charges, which influence the ion transport and other diffusion processes, causing polarization at local corrosion elements. The mechanical stability of such membranes is limited and drying periods destroy them. As a consequence of the formation and rupture of these colloidal membranes the corrosion process becomes very erratic, which is illustrated by usually poor reproducibility of potential/time relationships and other electrochemical data.[35] It is difficult to imitate such erratic processes by suitable equivalent conductive circuits as used in the evaluation of EIS data.

A very important question in the evaluation of
conventional corrosion tests as well as of EIS data is
whether defects are characteristic of the coating system
or just a consequence of poor application. As all these
tests have a classifactory character, this question can
only be answered reliably by using a sufficient number of
samples to allow a statistical assessment.

Finally it must be considered that organic coatings,
especially freshly prepared ones, but also some time later
on, contain low molecular weight, volatile or water soluble
material, which may influence EIS if the samples have not
been exposed to water before. However, this is a
principal problem in testing organic coatings. Rammelt
and Reinhard[36] recently distinguished two typical limiting
cases of impedance/frequency relationships for organic
coatings exposed to aqueous media: coating systems with
local defects which are due to the application (pore
model), and coating systems with homogeneous permeation of
water, followed by underrusting, which illustrate a
system-typical behaviour (multilayer model) (Figure 1).

3 EXPERIMENTAL RESULTS IN TESTING ORGANIC COATINGS BY EIS

In order to check this classification and to obtain more
information on the usefulness of EIS for evaluating and
predicting the corrosion protective quality, different
unpigmented coatings were studied.[37] The coatings were
applied to clean unpretreated steel panels with a dry film
thickness of 30 ± 2µm. These samples were exposed to 0.1M
aqueous solutions of NaCl for both EIS and wet adhesion
measurement by the adhesive tape test.

Ideally, intact coatings are expected to show pure
capacitative behaviour with a slope of -1 in the
impedance/frequency diagram (Bode-diagram). Figure 2 is a
Bode-diagram of an organic coating which showed poor wet
adhesion and a few tiny rust spots but still no underrusting. As the slope of the line in the Bode-diagram
indicates pure capacitative behaviour, it must be
concluded that the ionic transport through the film is
negligible. The visible rust spots had obviously been
blocked by corrosion products at the time of measurement.
That wet adhesion had strongly decreased, was not
detectable by EIS as probably no ions could penetrate the

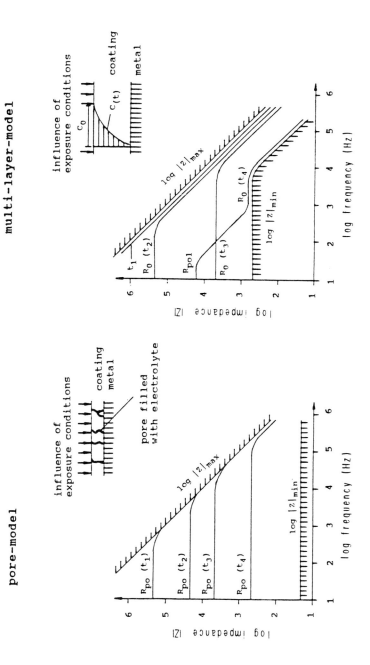

Figure 1 Schematical Bode-diagrams for coatings with local (pores) and extended (underrusting) failures resp.

(U. Rammelt, G. Reinhard – Farbe + Lack 98 (1992) 261)

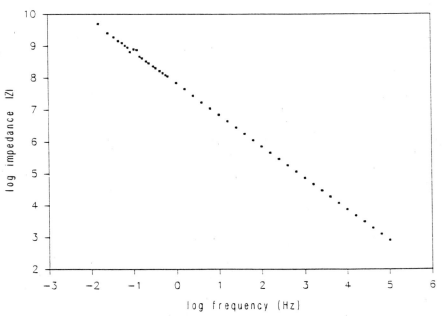

Figure 2 Bode-diagram of an organic coating with poor wet adhesion but no underrusting

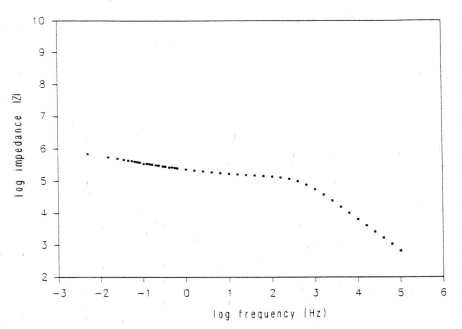

Figure 3 Bode-diagram of an organic coating with defects and underrusting

film and no underrusting occurred. Therefore one has to
be careful in using the Bode-diagram, only, to decide
whether good protection by a coating can be expected.

An organic coating with defects, blisters and
underrusting is clearly recognized by a strong decrease of
the complex impedance toward the low frequency part of the
curve (Figure 3). Such a diagram is very typical for
defective, unsuitable coating systems.

Corrosion processes at coating defects as well as
under coatings (underrusting) are erratic and dynamic by
nature. It is therefore important to pursue the
impedance/frequency relationships over a longer time. In
Figure 4 three Bode-diagrams have been obtained after 1
hour, 7 days and 14 days respectively during exposure to
0.1M aqueous solutions of NaCl. The curves do not differ
very much except by a small deviation of the slope in the
low frequency range of the second diagram (7 days). The
frequency dependency of the phase angle indicates this
deviation more distinctly. However, after 14 days
exposure, the curve is identical to that after 1 hour
exposure. Very obviously some corrosion occurred after 7
days, but later on the ionic pathways were blocked by
corrosion products which prevented ion transport, at least
temporarily, feigning good protection despite the fact
that visual inspection indicated rust spots and blisters.

In order to find out how reliably EIS measurements
with a single sample may evaluate the corrosion protective
quality, 17 samples have been prepared in the same way,
taking care that all parameters of composition, application and film formation were kept as constant as possible.
These samples were exposed to 0.1M aqueous solutions of
NaCl and the time of exposure was noted when the ohmic
resistance dropped below $R_{po} = 10^{10}\Omega$. As seen in Figure 5,
seven samples out of 17 failed to keep high electrical
resistance longer than two days and only two samples
remained intact after 504 hours of exposure.

The conclusion to be drawn from this experiment was
that these two samples were representative of the
protective quality of the coating film, whereas the
residual 15 samples had defects originating from their
preparation which led to failures after different times
according to the severity of the defect. This points out

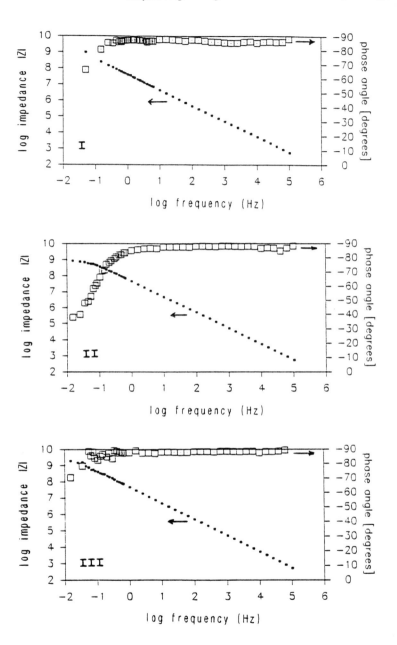

<u>Figure 4</u> Bode-diagrams of a sample with defects and underrusting after (a) 1h, (b) 7d and (c) 14d of exposure in a 0.1M NaCl-solution

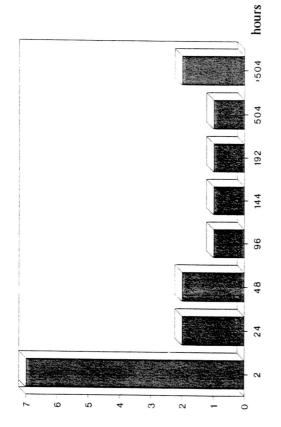

Figure 5 Time of exposure in water [h] of 17 samples after which the electrical resistance has decreased below $R_{p0} = 10^{10}\,\Omega$

how carefully EIS results have to be considered to avoid misinterpretation.

4 CONCLUSION

Electrical impedance spectroscopy is a very sensitive method to detect defects in organic coatings which are large enough not to be blocked by corrosion products. Ion conductivity is indicated by a drop of the polarization resistance, starting at the low frequency end of the EI spectra. A decrease of wet adhesion of organic coatings may remain undetected by EIS as long as no ion transport occurs and no underrusting takes place.

Systems with pure capacitative behaviour may still differ in their protective quality, especially if the exposure conditions are different from those of the EIS experiment. The reproducibility of EI measurements, which depends on the sample preparation, should be secured. Therefore the statistical relevance of the results has to be checked by testing a larger number of samples. Only then is it possible to distinguish between the system-specific and the application-dependent protective quality of organic coatings.

ACKNOWLEDGEMENT

The support of the Arbeitsgemeinschaft für industrielle Forschung and the Bundesministerium für Wirtschaft is gratefully acknowledged.

REFERENCES

1. F. Tödt, "Korrosion und Korrosionsschutz", Walter de Gruyter, Berlin, 1955, p.1014.
2. J.H. Wirth, Chem Fabrik, 1938, 11, 455; Korrosion und Metallschutz, 1940, 16, 69 and 331; Angewandte Chemie, 1941, 54, 369.
3. F. Wormwell and D.M. Brasher, J. Iron Steel Inst., 1949, 162, 129.
4. R.Ch. Bacon, J.J. Smith and F.M. Rugg, Ind. Eng. Chem., 1948, 40, 161.
5. E. Patrick, Dissertation Universität Stuttgart, 1978.
6. H. Leidheiser, Progress Org. Coatings, 1979, 7, 79.
7. G. Reinhardt and K. Hahn, Plaste und Kautschuk, 1975, 22, 361.

8. G. Menges and W. Schneider, Kunststofftechnik, 1973, 12, 265, 316, 343.
9. H. Potente and F. Stoll, Farbe & Lack, 1975, 81, 701.
10. F. Mansfeld, M.W. Kendig and S. Tai, Corrosion, 1982, 38, 9.
11. T. Szauer, Progress Org. Coatings, 1982, 10, 171.
12. H. Corti and R. Fernandez-Prini, Progess Org. Coatings, 1982, 10, 5.
13. M. Piens, R. Verbist and J. Vereecken, Proc. IX Internat. Conf. I. Org. Coatings and Science Technology, Athens, July 1983.
14. F.M. Geenen, Dissertation Techn. Universität Delft, 1991, Pasmans Offsed Druckerij, B.V. Gravenhage.
15. U. Rammelt, Habil. Schrift Techn. Universität Dresden, 1991.
16. J. Ross and J.R. MacDonald, "Impedance Spectroscopy", J. Wiley and Sons, New York, 1987.
17. J. Wolstenholme, Corrosion Science, 1973, 13, 521.
18. G. Reinhard and K. Hahn, Plaste und Kautschuk, 1979, 26, 580.
19. H.F. Clay, J. Oil Col. Chem. Assoc., 1969, 52, 158.
20. J.D. Scantlebury and K.N. Ho, J. Oil Col. Chem. Assoc., 1979, 62, 82.
21. W. Funke and H. Zatloukal, Farbe & Lack, 1980, 86, 870.
22. F.L. Floyd, R.G. Groseclose and C.M. Frey, J. Oil Col. Chem. Assoc., 1983, 66, 353.
23. D.J. Mills and J.E.O. Mayne, J. Oil Col. Chem. Assoc., 1984, 67, 49.
24. W. Funke, Pitture e Vernici, 1984, 7, 42.
25. B.R. Appleman and P.G. Campbell, J. Coatings Technol., 1982, 54(686), 17.
26. R.D. Wyvill, Metal Finishing, January 1982, 21.
27. D. Saatweber, Farbe & Lack, 1981, 87, 190.
28. D. Saatweber, Surtec Berlin, 1983, VDE-Verlag GmbH, Vol C8, p.131.
29. K. Lampe and A. Saarnak, Scandanavian Paint and Printing Ink Res. Institute, NIF Report T9-32M, July 1982.
30. T.R. Bullett, "Färch och Lack", 1983, 122.
31. W.A. Higgins, Cleveland Soc. Paint Technology, Official Dig., 1961, 37(490), 1.
32. T. Misewa, K. Hashimoto and S. Shimodaira, Corrosion Science, 1974, 14, 131.
33. T. Misewa, K. Asami, K. Hahimoto and S. Shimodaira, Corrosion Science, 1974, 14, 279.

34. R. Grauer, <u>Werkstoffe und Korrosion</u>, 1969, <u>20</u>, 991.
35. J.C. Galvan, S. Feliu and M. Morcillo, <u>Progess Org. Coatings</u>, 1989, <u>17</u>, 135.
36. U. Rammelt and G. Reinhard, <u>Farbe & Lack</u>, 1992, <u>98</u>, 261.
37. J. Prause, Diss. Universität Stuttgart, 1992.

1.4.2
Thin Film Electro-polymers as Organic Conversion Coatings

J. Marsh, J. D. Scantlebury, and S. B. Lyon

CORROSION AND PROTECTION CENTRE, UMIST, PO BOX 88,
MANCHESTER M60 IQD, UK

1 INTRODUCTION

<u>Non-pigmented Organic Coatings and Corrosion Protection.</u>

The mechanism of protection of an iron substrate by clear lacquer organic coatings was investigated by Mayne[1]. It was concluded that coatings are so permeable to the diffusion of water and oxygen that these are not the limiting factor in determining the corrosion rate. Clear lacquer films are in fact believed to prevent corrosion by creating an area of high ionic resistance, impeding the movement of ions to and from the metal surface and thus hindering the anodic and cathodic reactions. The importance of ionic movement in the corrosion of coated metals has been identified experimentally by Mayne[2,3], using organic films containing polarisable groups or areas of low cross-linking density.

A recent area of discussion has arisen within this concept, based on the importance of coating adhesion with respect to corrosion resistance, as opposed to bulk ionic migration through the coating. This is important because if interfacial breakdown and interfacial ionic movement is the dominant factor in the corrosion mechanism, an increase in the wet adhesion should lead to a decrease in the substrate corrosion rate. The main proponent of this theory is Funke[4,5].

However, if bulk ionic migration is dominant, then the substrate corrosion rate should be more or less independent of the adhesion of the coating, assuming a basic level of adhesion capable of maintaining the coating in contact with the substrate. This has been examined experimentally by Walker[6], who found little or no correlation between coating adhesion and coating resistance, with systems possessing similar wet adhesion values, such as polyurethane and alkyd coating systems, resulting in massively different corrosion rates. Work by Costa and Scantlebury[7] on alkyd coated steel has shown the importance of coating flaws with respect to corrosion, again indicative that bulk ionic transfer phenomena are of more importance than coating adhesion.

The Electro-Polymerisation of Phenol Derivatives.

Electro-polymerisation is the polymerisation of a monomer system directly onto an electrode from solution, via an electro-oxidative or electro-reductive process initiated by a suitable applied voltage.

The electro-polymerisation of phenol derivatives occurs on a large number of metals, including platinum, iron, copper and nickel[8,9], and is an electro-oxidative process based on the production of neutral phenoxy radicals[10] or electron deficient phenoxonium cations[11] via a one (for radicals) or two (for cations) electron oxidation. Polymer formation occurs due to coupling between radical mesomers, or via nucleophilic-electrophilic coupling of the phenoxonium cations with phenoxy anions or phenol molecules. The three possible oxidation states for phenol are shown in Figure 1.

Although the author has investigated the electro-chemistry of this reaction in depth, the important note to be made for the purposes of this paper is that both reactions should produce the same keto-intermediate dimer, which tautomerises to regenerate a phenolic group. Further reactions can then occur, forming a polyphenylene-oxide structure as shown in Figure 2. This has been identified as the product of practical electro-polymerisation experiments by several workers, via the use of infra red spectroscopy.

It has been shown by Dubois[12] that reactive functionality such as hydroxyl and amine groups present in the monomeric phenol molecule can be retained and can react with chemical systems applied to the polymer surface. Musiani and Mengoli[13] have produced experimental evidence showing that double bond functionality is also retained, and that cross-linking and functionality loss occurs on heat curing.

2   EXPERIMENTAL AND RESULTS

Specimen Preparation

The steel specimens to be coated were standard Q-panels, which were prepared by polishing the ground surface using 320 grit silicon carbide papers, then cut in half across the breadth of the panel and degreased with methanol. The panels were then prepared in a number of ways.

Control panel preparation. The control panels were coated with alkyd using an insulation tape well and a glass rod (dry thickness 13-20μm, Elcometer induced current thickness meter) on the ground surface and the edges were sealed using insulation tape. The panels were then cured at 55°C for 4 days in the presence of oxygen to produce a hard cured coating. The rear of the panels was then sealed using 3/1 bees wax/ calophony resin.

Reactive filmed panel preparation. The reactive panels were prepared by polarising the panels at +1000mV (SCE) for 10 minutes in a .25M solution of 2-allylphenol in

Table 1   Time (in days) of 50% specimen corrosion
           initiation for alkyd coated steel with various
           steel pretreatments.

No treatment   3 days

Reactive film   5 days

Non-reactive film   3 days

Silanisation   3 days

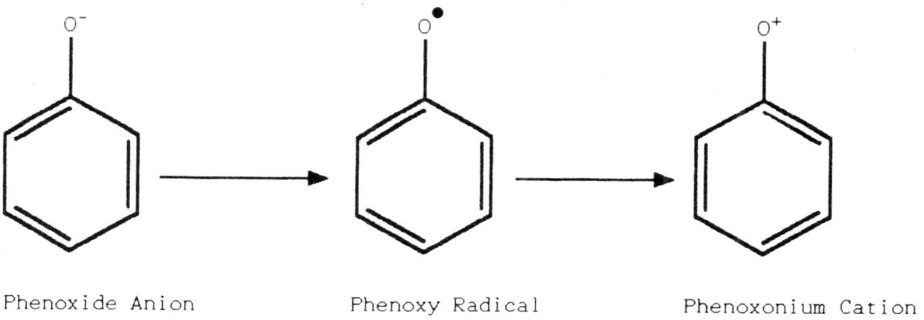

Phenoxide Anion   Phenoxy Radical   Phenoxonium Cation

Figure 1   The three possible oxidation states of phenol

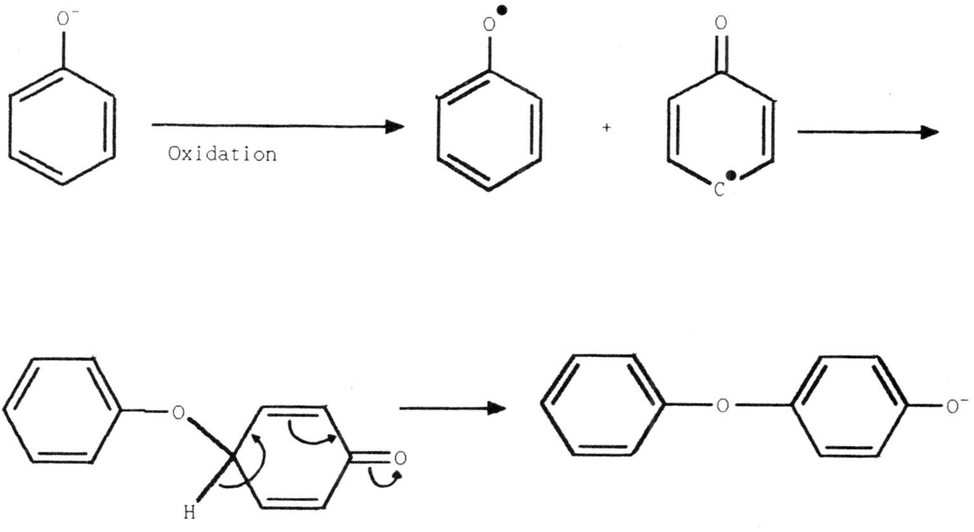

Figure 2   Polymerisation mechanism. (Radicals)

10/10/1 deionised water /methanol/2-ethoxyethanol, pH 10.2,
using allylamine as the base. This is a system known to
produce thin (< 0.5μm), highly adherent polyphenylene oxide
films   possessing   allylic   functionality   on   a   steel
substrate, and thus   reactive   compatibility with the alkyd
coating system. A second Q-panel was used as the cathode.
The panels were then dried in a vacuum desiccator. Coating
with alkyd was then performed as for the control system.

    Non-reactive filmed panel preparation.   These were
prepared as per the reactive panels, but the polyphenylene-
oxide film was heat cured for 12 hours at 90°C to greatly
reduce the level of allylic functionality present. The
panels were then coated as per the control system.

    Silanised panel preparation.   These were prepared by
immersing   the   panels   in   a   2%   v/v   solution   of
vinyltrimethoxysilane   in   methanol   for   5   minutes.   The
panels were then prepared as per the control system.

Immersion Tests

    Four panels were taken from each of the systems for
dry adhesion testing. A further 28 panels were immersed in
distilled water. Four panels were removed after 24 hours
and examined for signs of corrosion. The panels were then
tested for adhesive strength using the H.A.T.E. (hydraulic
adhesion testing equipment, Dunlop/Conoco) dolly pull off
test (a direct tensile test), in which 5 tests per panel
were performed, and the mean value for adhesive strength
obtained. The results of this test are given in Figure 3.
The ASTM standard cross hatch test was also used, with a
total of 11 tests performed and the median value taken. The
results of this test are expressed in Figure 4. Four panels
were removed at further 24 hour periods and the corrosion
and adhesion properties examined as above. The total length
of the immersion test was thus seven days. The period of
time occurring before significant initiation of corrosion
was noted on the majority of specimens for each system is
expressed in Table 1. A microscopic comparison of the
corrosion present at 2 days immersion was made between the
reactive filmed system and the control system, with optical
micrographs shown in Figures 5 and 6.

3   DISCUSSION

Poly(2-allyl)phenylene-oxide films as adhesion enhancers.

    The   hypothesis   for   this   experiment   is   that   the
accepted curing mechanism of alkyd coatings is based on the
formation of a radical or peroxy group at the methylene
group adjacent to the C=C groups present in such systems,
these radicals or peroxy structures then reacting with a
C=C bond present in an adjacent monomer molecule to produce
C-C cross-linking. The allyl groups present in the electro-
polymer film should interact chemically with this curing
mechanism, producing chemical linking to the alkyd coating.
This will lead to an adhesion enhancement if the adhesion
of the electro-polymer film to the steel substrate is

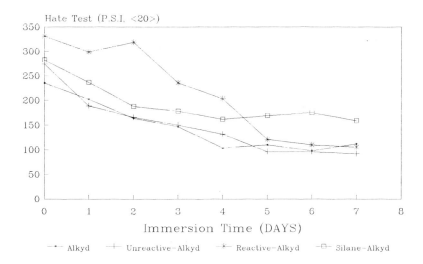

Figure 3 Hydraulic adhesion (H.A.T.E.) test results for alkyd coated steel with no treatment, reactive polyphenylene-oxide, unreactive polyphenylene-oxide and silanisation surface pretreatments. Water immersion time 0-7 days

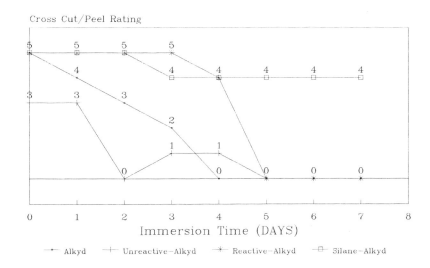

Figure 4 ASTM cross-hatch test results for alkyd coated steel with no treatment, reactive polyphenylene-oxide, unreactive polyphenylene-oxide and silanisation surface pretreatments. Water immersion time 0-7 days

Figure 5   Optical micrograph - 15μ alkyd coated steel with reactive film treatment, 48 hours immersion in deionised water

Figure 6   Optical micrograph - 15μ alkyd coated steel with no pretreatment, 48 hours immersion in deionised water

greater than that of the alkyd coating to the steel. We can
see from Figures 3 and 4 the results for alkyd coated steel
and alkyd/reactive film coated steel.

A significant adhesion enhancement does occur, both
for dry conditions and up to 4 days immersion in distilled
water. After this period the adhesion falls to that
observed with no thin film present, showing that any
adhesion enhancement is relatively short term with respect
to continuous immersion conditions.

Examining the results for low-reactivity filmed/alkyd
coated steel, no adhesion enhancement can be observed using
the H.A.T.E test, while the performance using the cross-
hatch test is worse than for the control system. Failure
occurs almost exclusively at the film/coating interface,
indicating that the bonding across this interface is
severely weakened by the loss of reactive compatibility
with the alkyd coating. The adhesion enhancement observed
for the reactive film system can thus be directly related
to the chemical interaction stated previously.

Comparing these results to silanised steel it can be
observed that the reactive filmed steel has better short
term adhesive strength under immersion conditions, but that
as immersion time increases, silanisation appears to become
the superior treatment for coating adhesion. Both are a
significant improvement on no treatment or a low reactivity
treatment.

Poly(2-allyl)phenylene-oxide   films   for   corrosion resistance.

Examining Table 1 it can be seen that the onset of
visually detectable corrosion is restricted only when the
reactive film is present. Microscopic examination of the
reactive polyphenylene-oxide system at 2 days (Figure 5)
immersion indicates that no detectable substrate corrosion
has occurred, compared to the control system (Figure 6)
where extensive micro-corrosion and micro-blistering can be
observed. This at first appears to indicate that superior
adhesion has lead to improved corrosion resistance,
supporting the interfacial breakdown theory. However, two
arguments complicate this. Firstly, no improvement in
corrosion resistance can be observed for the low reactivity
filmed steel system, even though the interfacial adhesion
between the steel and the thin film should be virtually
identical to that seen for the reactive filmed system.
Noting that the polyphenylene-oxide film has a measurable
thickness, it is appears to be the adhesion of the bulk
coating, not the steel interfacial adhesion, that
predominantly affects the corrosion rate. Also although
silanisation leads to effective long term adhesion
enhancement, no effect on the onset or severity of under-
coating corrosion was observed in comparison to the control
system. A direct relationship between interfacial adhesion
enhancement and corrosion resistance cannot be stated.

4   CONCLUSIONS

This paper has been concerned with a new form of surface treatment, electropolymerisation, for adhesion enhancement and corrosion resistance with respect to coated steel. The results clearly show significant enhancement of both of these properties when using a monomer system containing functionality compatible with the curing scheme of the top-coating system. With respect to short term immersion this enhancement is clearly superior to a compatible silanisation system. Over longer term immersion silanisation becomes the superior form of surface treatment with respect to adhesion enhancement, but throughout the experiment no reduction of corrosion compared to the control system could be observed, as opposed to the reactive film treatment where a significant reduction in corrosion occurred.

With respect to a mechanistic interpretation of corrosion under organic coatings, these results seem to complicate the situation with respect to the theories stated in the introduction. Adhesion clearly has an effect in the filmed steel systems, but this cannot be equated with the results observed using silanisation. The indication may be that while adhesion is important for corrosion prevention during short periods of immersion in or exposure to water, corrosion due to longer term immersion may be independent of the coating adhesion, a conclusion also reached by Walker[6].

REFERENCES

1.   J.E.O. Mayne, "Corrosion Vol 2, Corrosion Control", Chapter 15.3, ed. L.L. Shreir, Newnes-Butterworths, London, 1978.
2.   J.E.O. Mayne, JOCCA., 1949, 32, 481.
3.   "   "   1951, 34, 473.
4.   W. Funke, extended abstracts-"Advances in Corrosion Protection by Organic Coatings", Cambridge, 1989, 35.
5.   H. Leidheiser, W. Funke, JOCCA., 1987, 70, 121.
6.   P. Walker, Off. Dig., 1965, 1561.
7.   I. Costa, Ph.D Thesis, UMIST, 1991.
8.   J. Marsh, MSc Dissertation, UMIST, 1989.
9.   G. Mengoli and M.M Musiani, J. Electrochem. Soc., 1987, 643C.
10. M.C. Pham, P.C. Lacaze, J.E. Dubois, J. Electroanal. Chem., 1978, 86, 147.
11. A. Ronlan, V.D. Parker, J. Chem. Soc. (C). 1971, 3214.
12. J.E. Dubois, P.C. Lacaze, M.C. Pham, J. Electroanal. Chem., 1981, 117, 233.
13. G. Mengoli, M.M. Musiani, F. Furrlanetto, proceedings, "Advances in Corrosion Protection by Organic Coatings", ed. J.D. Scantlebury, M.W. Kendig, Electrochem. Soc., Pennington, New Jersey, 1989, 198.

1.4.3
The Structure and Properties of Plasma Sprayed Polyamide 11 Coatings

Y. Bao and D. T. Gawne

DEPARTMENT OF MATERIALS TECHNOLOGY, BRUNEL, THE
UNIVERSITY OF WEST LONDON, UXBRIDGE, MIDDLESEX, UK

1   INTRODUCTION

Polyamide 11 is applied as a protective coating on a wide range of
engineering components (1) mainly by electrostatic spraying and
fluidized bed dipping. Plasma spraying has potential as an
alternative coating process because its high velocity,
inert/reducing plasma is expected to reduce degradation and
porosity, while promoting adhesion. The process involves injecting
solid particles into a plasma jet, where they are heated,
accelerated and projected on to a substrate to form a coating (2).

This paper concerns depositing polyamide 11 coatings on plain
carbon steel over a range of arc power levels and investigating
their structure and mechanical properties. The work is directed at
identifying the critical flaws in the coatings, their formation
mechanisms, and influence on tensile properties and wear behaviour.

2   EXPERIMENTAL PROCEDURE

The coating powder used was polyamide 11 (Rilsan BHURX grade)
supplied by Atochem UK Ltd.. Scanning electron microscopy revealed
an equiaxed angular particle morphology and a mean size of 80μm.
The substrate used was a plain carbon steel (080M40 grade) in the
form of a plate of 6mm thickness. The steel was degreased and grit
blasted with Metcolite C grit with a blast pressure of 4 bar and a
blast distance of 150mm to give a surface roughness R_a value of 7μm.

Plasma spraying was undertaken using a Metco plasma spray
system with a MBN torch, MCN control unit, 4MP powder feed unit and
fluidized hopper. Polyamide 11 coatings of 500μm in thickness were
produced using a nitrogen-hydrogen plasma gas.

Wear performance was assessed using a reciprocating pin-on-flat
machine with two types of pin surfaces: a stainless steel ball and
Rockwell diamond. The stainless steel ball was of diameter 12.7mm
and the Rockwell diamond consisted of a regular cone with a 120
degree included angle. Loads of 20N and 4N were applied to the
steel ball and diamond respectively, which slide at 50 cycles per

minute over a track length of 30mm on the flat coated plate. Wear
was evaluated by measuring the mean depth of the wear track on the
polymer coating using a linear variable differential transducer.

Tensile testing was undertaken on free-standing coatings of
gauge length 16mm using an Instron TT-M machine with a cross-head
speed of 50mm min^{-1}. The structure of the coatings were examined by
light microscopy and scanning electron microscopy. Cross-sections
of coatings for microstructural examination were prepared by
separating the coating from the substrate, embedding in Araldite,
polishing and etching in nitric acid at room temperature.

3 RESULTS AND DISCUSSION

(a) Arc Power

Coatings were plasma sprayed at several arc power levels
between 15kW and 30kW using an nitrogen-hydrogen plasma gas with a
torch traverse speed of 150mm s^{-1}. Figure 1 shows the results from
subsequent wear testing. Both wear measurement techniques indicate
an optimal arc power of 21kW for minimum wear rates with the
coatings exhibiting high wear rates at lower and higher power
levels.

(b) Structure

Examination of through-thickness sections of the plasma-sprayed
coatings revealed that the polyamide deposits were composed of
elongated regions with the major axes parallel to the coating-
substrate interface (Figure 2a), whereas the top surface parallel to
the interface consisted of equiaxed regions (Figure 2b) with similar
diameters to the major axes of the regions in the through-thickness
section. The deposit is therefore composed of disc-like entities or
splats resulting from the high-velocity impact of the molten
spheroidal droplets from the plasma stream on to the substrate. The
aspect ratio of the splats is typically four indicating that
considerable flow occurs on impact. There is a significant
variation in aspect ratio and although this is partly due to the
geometrical effect of sectioning discs through their thicknesses, it
is also considered to be a consequence of differing degrees of flow
of the molten particles impinging the substrate. Since the powder
composition and intrinsic viscosity of the polyamide powder was
constant, the variation in droplet flow is likely to be due to
differing temperatures attained by the particles in the plasma jet.
This can arise from a number of reasons including a non-ideal
particle trajectories in the plasma jet or too large a particle size
resulting in insufficient time for thermal conduction to its centre.
In extreme cases, this can lead to incomplete melting of the coating
powder particles in the plasma. The feature towards the top right-
hand side of Figure 3 is an inadequately melted particle exhibiting
poor flow and consequent voids around its periphery. The void in
the centre of Figure 3 is attributed to an unmelted particle, which
was subsequently pulled out during polishing for microstructural
examination.

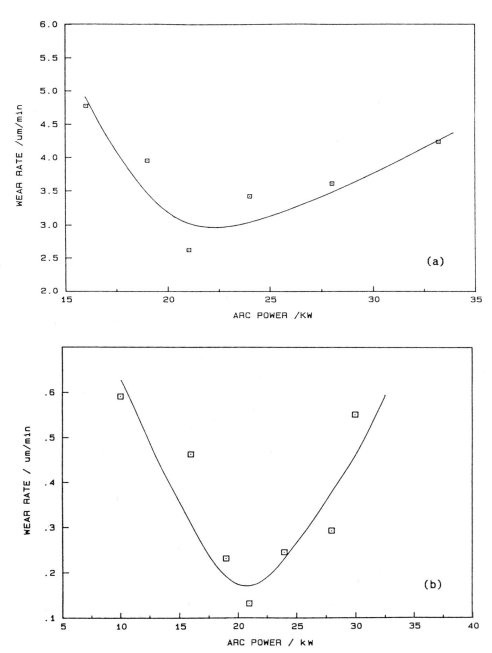

<u>Figure 1</u>   Effect of arc power on the wear rate of plasma sprayed
         polyamide 11 coatings against (a) stainless steel ball,
         and (b) Rockwell diamond

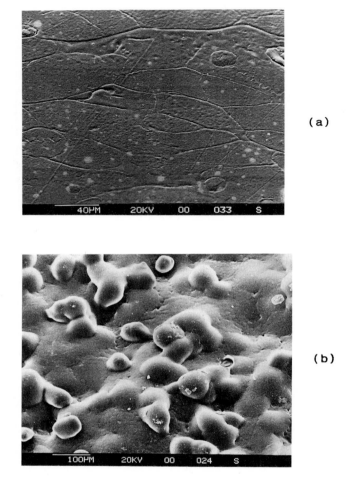

Figure 2   Scanning electron micrographs of polyamide 11 coating: (a) through–thickness section and (b) top surface parallel to interface

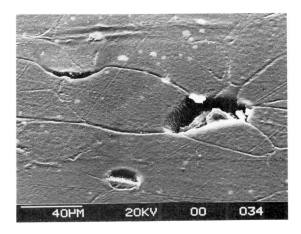

Figure 3 SEM of through-thickness section of a polyamide 11 coating showing a partially melted particle and voids

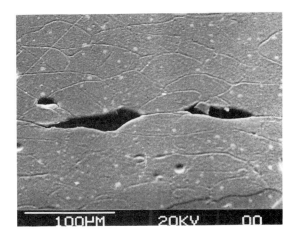

Figure 4 SEM of through-thickness section of a polyamide 11 coating showing a gas entrapment between splats

Incomplete melting resulting in insufficiently low viscosity and particle flow on impact with the substrate therefore has two deleterious effects: void formation due to lack of conformity with neighbouring splats and weak inter-splat bonding as evidenced by particle pull-out on light polishing.

Figure 4 shows long elongated voids between neighbouring splats suggesting gas entrapment between the coating particles during spraying. The high aspect ratio and sharp curvature of these voids, together with their alignment along the splat boundaries would indicate a particularly adverse effect on mechanical properties under tensile stress systems.

The fourth type of void observed in the coating was spherical and located in the splat interior away from the splat boundaries as shown in Figure 5. To gain further information on this phenomenon, wipe tests were carried out in which the plasma torch was traversed extremely rapidly over the substrate so that isolated splats were produced. Subsequent examination under the SEM showed that voids were evident in some individual splats (Figure 6). The incidence of these voids increased with increasing arc power. The results suggest that these voids are formed by thermal degradation of the polymer to generate gaseous products.

(c)   Tensile properties

The coating deposited under a sub-optimal arc power of 18kW exhibited a tensile strength of 20MNm^{-2} and a total elongation to fracture of 9%, whereas the coating sprayed at the optimal power of 21kW gave a tensile strength of 40 MNm^{-2} and total elongation of 25%. Figure 7 shows a micrograph of the through-thickness fracture surface from the tensile specimen of the 18kW coating. The fracture surface is predominantly brittle in nature. However, the fracture surface of the 21kW arc power coating displays a ductile appearance (Figure 8) in marked contrast to the latter power coating. These results indicate that fracture in the sub-optimal coatings occurs along the splat boundaries due to stress concentrations at voids and weak inter-splat bonding. However, there is sufficient particle melting and strong bonding between splats in the optimal coating to allow stress transfer across the splat boundaries and extensive plasticity (permanent, irreversible deformation) of the polyamide to take place. Figure 9 shows that the optimal coating exhibits substantially more tensile ductility than the sub-optimal coating in support of this view.

(d)   Tribological properties

Figure 10 shows the progress of sliding wear of polyamide 11 coatings against a stainless steel ball. An initial period of rapid wear or running-in was observed over the first 500 cycles followed by a more gradual equilibrium wear period. A similar pattern of behaviour was observed under reciprocating sliding wear with the Rockwell diamond stylus as shown in Figure 11. The initial high wear period is attributed to the removal of protrusions and high spots on the original surface, and misalignments between the sliding surfaces.

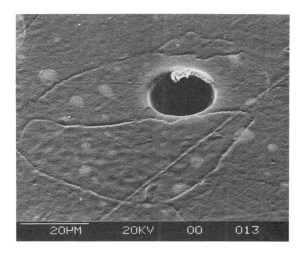

Figure 5   Spheroial void in the interior of a splat in a polyamide 11
            coating

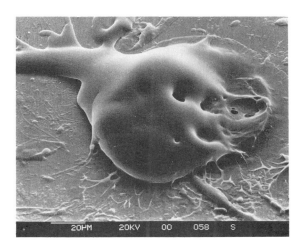

Figure 6   Voids within an individual wipe test particle of polyamide
            11 powder

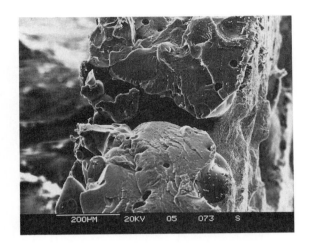

Figure 7   Fracture surface of polyamide 11 coating sprayed at 18kW arc power

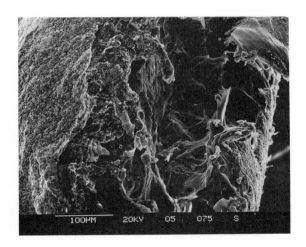

Figure 8   Fracture surface of polyamide 11 coating sprayed at 21kW arc power

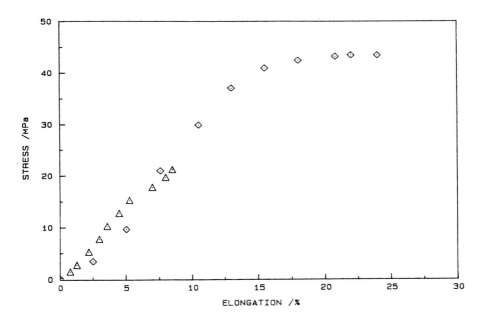

<u>Figure 9</u> Stress-strain curves for optimal (◊ 21kW)
          coating and sub-optimal (▲ 18kW) coating

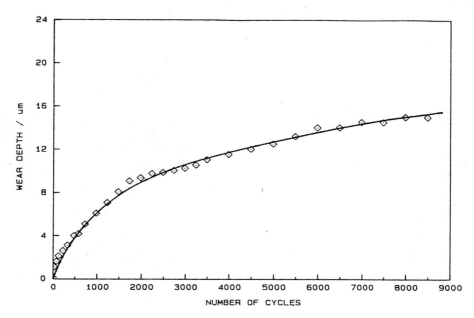

Figure 10   **Wear of plasma deposited polyamide 11 coatings against
a stainless steel ball**

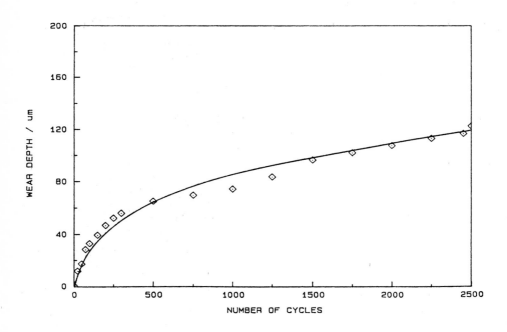

Figure 11   **Wear of plasma deposited polyamide 11 coatings against
a Rockwell diamond stylus**

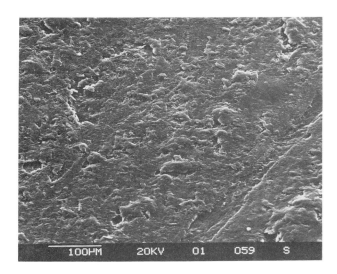

Figure 12   Polyamide 11 coating surface after wear against the
            stainless steel ball

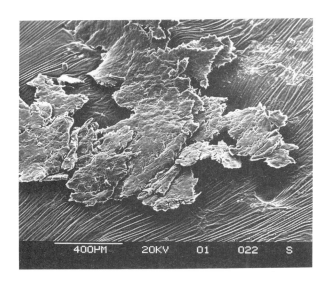

Figure 13   Wear debris from sliding of a polyamide 11 coating against
            the stainless steel ball

Figure 12 is a scanning electron micrograph from the sliding surface of the polyamide coating after wear against the stainless steel ball. The worn surface contains shallow pits which appear to have been caused by material being pulled out of the surface. Examination of the steel and diamond counterfaces revealed evidence of polyamide transfer films, suggesting adhesive wear. The debris particles as shown in Figure 13 were generally flat platelets of polyamide of similar dimensions to the splats making up the deposit. It is likely therefore that the polyamide adheres to the sliding counterface and fractures occur along the underlying splat boundary to generate a debris particle. The presence of voids in the coating were associated with enhanced local damage as shown in Figure 14 to the detriment of wear resistance. Figure 15 shows the relationship between porosity and wear rate, which is the practical result of these interactions. Figure 12 reveals the presence of grooves parallel to the sliding direction which is indicative of abrasive wear caused by asperities from the harder counterface ploughing out the softer polyamide surface.

The observed wear behaviour of the polyamide coatings consists of plastic deformation followed by fracture to produce debris particles. The presence of voids and bonding between splats play a crucial role in wear performance, and they can be influenced by controlling the plasma processing to ensure adequate particle melting while avoiding excessive degradation.

4   CONCLUSIONS

1.   Maximum wear resistance of polyamide 11 coatings on steel was achieved by plasma spraying with an arc power of 21kW under the conditions used.

2.   The plasma sprayed deposits are composed of overlapping disc-shaped splats. There are significant variations in the extent of flow of the particles on impact with the substrate and their degree of melting.

3.   Incompletely melted particles and voids are significant defects within the coatings and substantially affect the deposit properties.

4.   Four major types of void were observed in the deposits: (a) those associated with unmelted particles, (b) those resulting from inadequate splat flow, (c) those caused by gas entrapment, and (d) those resulting from thermal degradation of the polyamide during deposition.

5.   The porosity adversely affects the tensile strength of polyamide coatings with the high porosity coatings exhibiting predominantly brittle fracture and the optimal, dense coatings showing substantial plasticity.

6.   The wear resistance of the coatings decreases markedly with increasing porosity. Adhesive transfer and abrasive wear are the principal wear mechanisms under the conditions used with fracture commonly occurring along the splat boundaries.

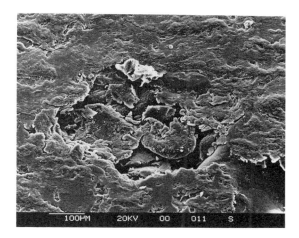

Figure 14   Wear track of polyamide 11 coating after sliding against the stainless steel ball

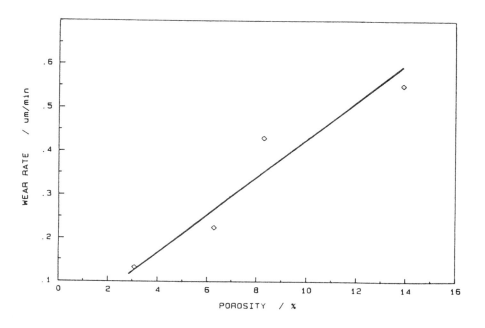

Figure 15   Effect of porosity on wear of polyamide 11 coating against the stainless steel ball

ACKNOWLEDGEMENTS

This work has been carried out with the support of the Procurement Executive Ministry of Defence, the Science and Engineering Research Council and Metco Ltd.. The authors would like to thank the above for permission to publish the paper.

REFERENCES

1.   C.E. Blackmore, Surface Engineering, 3 (1987) 29.

2.   A. Vardelle, M. Vardelle and P. Fauchais, Plasma Chemistry and Plasma Processing, 2 (1982) 255.

Section 1.5 Metallic Coatings

1.5.1
Surface Finishing and the Environment: Alternatives to Cadmium as a Surface Coating

D. R. Gabe

INSTITUTE OF POLYMER TECHNOLOGY AND MATERIALS
ENGINEERING, LOUGHBOROUGH UNIVERSITY OF TECHNOLOGY,
LOUGHBOROUGH, LEICS. LEI I 3TU, UK

1 INTRODUCTION

In this era of environmental consciousness all processes
and materials must be examined for their impact on the
environment. In particular the metallic elements have long
been classified in the context of food and nutrition and
their general toxicity known (see Table 1). Some of these,
by their very nature, are little used, for example the
metalloids arsenic, antimony and bismuth, while others are
used in relatively restricted manners, for example lead.
Some metals however continue to be used despite their known
toxic characteristics either because they are known to be
safe in certain circumstances or because they are gradually
being phased out over a longer time scale.

Several examples can be cited. The use of lead has
diminished in many ways but is still in use for batteries
and industrial electrodes because in such uses it is
essentially insoluble. Chromium is toxic in the hexavalent
state (ie. chromate or dichromate) but non-toxic in the
trivalent or metallic state from which it cannot easily be
oxidized back to the hexavalent state. When employed as a
protective coating the process solution is hexavalent but
the coating contains none as residue; thus when chromate
is used as a process solution its solution wastes are
easily reducible to safe trivalent forms; its product is a
passivated surface which is safe in a trivalent form and as
yet has no good alternative.
Cadmium is being replaced by alternatives and within ten
years will be little used for surface coatings. Amongst
the top ten coating metals not only is it one of the most
toxic but it is also perhaps the least vital or unique in
its usage options.

Table 1 Classification of trace elements in foods

Metal	Nutritive	Nonnutritive Nontoxic	Toxic
Cobalt	X		
Copper	X		
Iron	X		
Manganese	X		
Zinc	X		
Aluminium			X
Beryllium			X
Chromium		X	
Silicon		X	
Strontium		X	
Tin		X	
Titanium		X	
Nickel			X
Silver			X
Bismuth			X
Cadmium			X
Mercury			X
Lead			X
Arsenic			X
Antimony			X
Barium			X
Selenium			X
Tellurium			X
Molybdenum			X
Vanadium			X
Thallium			X

Table 2 USA Environmental Protection Agency regulations for metal-containing effluent

Pollutant	Metal Finishing Guidelines* Concn., ppm		Electroplating Guidelines** Concn., ppm	
	Daily Max.	Monthly Av.	Daily Max.	Max. over 4 days
Cyanide	1.2	0.65	1.9	1.0
Total suspended solids	60	31		
Cd	0.69	0.26		
Cd (new source)	0.11	0.07	1.2	0.7
Cr (total)	2.77	1.71	7.0	4.0
Cu	3.38	2.07	4.5	2.7
Pb	0.69	0.43	0.6	0.4
Ni	3.98	2.38	4.1	2.6
Zn	2.61	1.48	4.2	2.6
Ag	0.43	0.24	1.2	0.7
Au			1.2	0.7
Total Metals			10.5	6.8
Oil, grease	52	26		
Total toxic organics	2.13	-		

*EPA Regulation 40 CFR413
**EPA Regulation 40 CFR 433

Some other metals like aluminium and nickel have some very specific problems but are not being seriously attacked from the environmental point of view. Metals like tin and zinc are widely used as coatings and are essentially non-toxic unless the concentrations are unusually high.

Besides the metals some acid radicals and anions are affected. Cyanide has long been known as an extremely versatile complexing agent and while alternatives are known for some uses they themselves may be suspect and are usually markedly inferior. The virtue of using cyanide is that it is easily destroyed by oxidation with chlorine or hypochlorite and consequently requires careful technology and not high technology.

In this paper the position of cadmium in metal finishing will be examined and the range of alternatives discussed in detail.

2 THE 'EPA' DIMENSION

Curiously enough limitations on the use of specific metals arise not from the products and their environmental deterioration but on the processing and the consequences on the environment for the processing wastes. The reason is not difficult to understand. The engineering products are designed to be environmentally stable and therefore corrode relatively slowly even in the case of metals like cadmium which are used in sacrificial modes to protect steel substrates in the form of engineering components. Thus if a cadmium coating of 5µm thickness dissolves with a life of 20 years the aqueous corrosion product is only rarely likely to have a local metal concentration above 1-2 ppm and will frequently be well below it. By comparison waste process solutions, arising from pickling, cleaning or electroplating unit processes may well contain 1000 ppm as dragouts and 50 ppm as rinses.

The Environment Protection Agencies, led by that in the USA, are therefore legislating against process wastes but can demonstrate different levels of acceptability according to type of industry involved. The stringency of the level imposed varies from metal to metal but without any fundamental criterion apparently involved, and some metals are not listed either because they are considered to be non-toxic (eg. tin) or have insignificant usage (eg. antimony). In any case a total metal limit will ultimately provide some control.

The present American EPA regulatory levels are given in Table 2 and it can be seen that cadmium has the most severe restrictions of those metals cited. But this level while clearly stringent cannot be directly related to toxicity simply because the toxicity is so uncertain. Thus recently maximum allowable intakes of cadmium of 200-2000 µg/day have been quoted [1] while average dietary intakes have been measured at 12-50 μg/day : drinking water limits have been set at 5-10 μg/l which is a parallel level [2].

Have these levels for cadmium directly reduced its usage? Two points need to be made : firstly that 0.07 ppm as a monthly average discharge concentration can be achieved by <u>careful</u> conventional precipitation methods. Secondly, the average metal finisher is not yet a <u>careful</u> effluent processor and in general has foregone the privilege of processing with cadmium; the secondary effect of this is that cadmium processing is now limited to relatively few industrial units whose throughput is sufficient to make <u>careful</u> effluent treatment economically feasible. Thus cadmium will not be driven out of usage but kept to the vital applications and be processed by reasonably sized and equipped units. Recent changes in the pattern of usage demonstrate that electrodeposited cadmium coatings have halved in tonnages employed (Figure 1, at end of Chapter) while battery consumption has increased markedly, a field in which cadmium has a more unique niche.

3 THE USE OF CADMIUM

The toxicity of cadmium, in the context of surface coatings, has two distinct aspects. Firstly, the process itself in which cadmium is electroplated from a cyanide-type solution and which produces a significant effluent treatment problem. Secondly, the metal itself which can corrode in service giving rise to toxic corrosion products. In general the toxicity is controlled through placing a concentration limit on its occurrence in waste waters, a limit varying from 0.01 to 1 ppm but frequently about 0.1 ppm, and varying at present from country to country and even between industrial sectors.

The applications of cadmium relate to its performance as a corrosion-protective coating primarily for steel in marine conditions where its sacrificial action can be ensured. The doubt arises because the standard potentials for iron and cadmium are very close and suggest that cadmium will be base to iron:

$$E^\emptyset, \quad Cd^{2+}/Cd \quad\quad - 0.403 \text{ V}$$
$$E^\emptyset, \quad Fe^{2+}/Fe \quad\quad - 0.440 \text{ V}$$

In practice, in the presence of chlorides, cadmium will be usually sacrificial to iron but in the presence of some other ions such as sulphate, bicarbonate etc. the reverse will be true, as it will also be if the cadmium coating is well passivated [3].

Cadmium is not often used as an undercoat but in principle it could be used as 'strike' undercoat for zinc on steel by virtue of its alkaline cyanide solution formulation.

Because of the price of cadmium it is rarely used as a thick coating and the usual specifications quote three thickness ranges.

For examples, ASTM A165

    3.9 μm min. Indoor applications, mild corrosion
    7.5 μm min. Outdoor applications, moderate corrosion
    13 μm min. Outdoor applications, marine corrosion.

In practice these thicknesses are further influenced by the
use of cadmium on steel threaded fasteners in which mating
tolerance is vital; in this instance it is common to limit
the thickness to 7-8 μm. The mating capability is further
enhanced by the surface lubricity of cadmium and by the low
volume of the cadmium corrosion products which enable
screw-threaded fasteners to be easily undone after periods
in service. For clips, brackets etc. the heavier
thicknesses are in principle appropriate.

The chemical resistance of cadmium per se is not
particularly good so that its use is limited to protective
sacrificial action. Discussion of alternatives is not new
and has been the subject of many earlier reviews [4-6].
The toxicity of cadmium is not in dispute and arises
through ingestion of cadmium as dissolved species causing
acute poisoning and inhalation of fumes of cadmium oxide
caused by volatilization in hot plant or overspraying of a
paint.

4 PROCESS ALTERNATIVES

The cyanide solution has been used for electrodeposition of
cadmium since 1920 and it is based on an alkaline
formulation (pH 8-12) in which the cadmium is present as
$Cd(CN)^-_3$ complex ions. Excess 'free' cyanide is necessary to
stabilize the complex and alkali to prevent HCN gas being
produced by decomposition. A good flexible process results
from the alkaline nature of the electrolyte allowing even
poorly cleaned surfaces to be well-coated.

The main alternative is fluoborate based and operated at pH
1-4. This is itself expensive and the fluoborate is only
slightly less toxic than cyanide and is not usually easier
to deal with in effluent treatment. Sulphate and
sulphamate solutions are also possible but as with all acid
electrolytes offer inferior throwing power when compared
with the cyanide solutions.

Whichever solution is used passivation, post treatments are
generally employed and these involve dips or sprays in
chromate or chromic acid solutions which themselves give
rise to effluent problems. However, these will not be
considered here.

The occurrence of hydrogen embrittlement is another
phenomenon associated more with cadmium than other metals.
It arises largely with high tensile carbon steel fasteners
(bolts not nuts) which absorb hydrogen during the cathodic
electroplating process. The phenomenon is well-known and
should not be a cause of difficulty because the
specification of a post-plating heat treatment for such

components is virtually mandatory. Formerly 1 hour at
200°C was considered a good 'rule of thumb' thermal
treatment but nowadays up to 20 hr. may be specified for
complete quality assurance in relation to brittleness. The
temperature of 200°C should not be exceeded as a means of
accelerating hydrogen dispersal.

However, concern on this matter has led to 'mechanical' or
'shot peen' plating being developed in which no side-
reaction hydrogen is available to cause brittleness and
consequently no post treatment is needed. Consequently, a
more expensive unit process may become economic in the
larger context.

5 PRODUCT ALTERNATIVES

The property requirements of cadmium plated components can
be summarized as follows [4-9] :
> Good corrosion protection
> Low frictional characteristics for bolt 'lubricity'
> Low volume corrosion products
> Compatibility with aluminium alloys
> Good solderability.

In practice the requirement for solderability tends to
separate two possible groups of alternative coatings:

i.   Zinc or zinc alloy coatings which are non-solderable;
ii. Tin or tin alloy coatings which are solderable

While the thickness requirement of 5-15 μm remains, the
implication is that electrodeposition is the appropriate
coating process (ie. not hot dipping or spraying) although
mechanical plating is a possible alternative, as mentioned
previously. The usual direct comparison is between cadmium
and zinc and the following conclusions can be drawn.

i.   In rural environments cadmium and zinc can offer
      similar protection although zinc is usually
      superior.
ii.   In industrial atmospheres zinc is superior.
iii. In marine atmospheres zinc and cadmium are similar
      but in contact with sea water and salt spray cadmium
      is superior.

In general cadmium has poor corrosion resistance to many
chemicals and in petroleum and oil industry liquids.
Furthermore, it cannot be used in the food or
pharmaceutical industries for obvious reasons. Its high
temperature usage is limited too because with a melting
point as low as 320°C it inevitably sublimes or volatilizes
at temperatures above 200°C and this is usually considered
to be the maximum temperature at which hydrogen de-
embrittlement should be carried out.

The alternative coating types which may be considered are
as follows, each of which being enhanced in performance by
an appropriate conversion coating (passivation) treatment:

        Plain zinc
        Zinc-tin alloys   (80% tin)
        Zinc-nickel alloys   (10-15% nickel)
        Zinc-cobalt alloys   (0.4-1.0% cobalt)
        Zinc-iron alloys   (20-80% iron)
        Zinc-manganese alloys (40-60% manganese)
        Plain tin

As is by now clear no single property or test performance
can define a substitute adequately but neutral salt spray
tests are widely used as a first criterion. On this basis
the results in Table 3-4   place the simpler options in
some perspective and immediately reveal the important role
of passivation treatments and then indicate that zinc, on a
similar thickness basis, is inadequate as a substitute. In
neutral salt spray (Table   3) the appearance of white rust
and red rust are used as indicators of progress of
corrosion; the appearance of red rust indicates that
little protection is occurring which in the case of tin is
to be expected. The time at which white rust appears is a
measure of the corrosion resistance of the coating itself
and typical values for salt spray are:

        Zinc unpassivated   1-4 hours
        Zinc yellow passivated   200-240
        Zinc olive passivated   120-150
        Zinc black passivated   200-240
        Cadmium yellow passivated   > 240

The importance of the passivation treatment can be further
seen in Table 4 where the effect of the type of drying
employed may also be noted. This effect is usually
attributed to the fact that the passivation film is
hydrated and if caused to dry too quickly it cracks and
yields a less protective film. Clearly if the drying is
too slow process hold-ups occur and so a compromise has to
be achieved between processing speed and film integrity
attained.

Comparable data can be obtained using accelerated cyclical
humidity tests (Table 5). Yellow passivated zinc and
cadmium (both 3 μm thickness) give similar performance but
tin is noticeably inferior as rust appears in very short
times.

If the choice of alternative coating is widened to include
zinc alloys the improved performance attainable is readily
apparent and the results for red rust appearance in Table
6 substantiate this conclusion. It should be noted that
much of the data for zinc alloy performance is obtained
from the steel strip industry and not from the engineering
and component industry. Thus while the applications in
automotive bodywork or architectural and building cladding

Table 3 Corrosion performance in neutral salt spray

Coating System (3 μm metal)	48 Hours		96 Hours		240 Hours	
	White Rust	Red Rust	White Rust	Red Rust	White Rust	Red Rust
Cadmium unpassivated	2	1	3	2	3	3
Cadmium yellow passivate	1	1	1	1	1	1
Zinc unpassivated	3	1	3	2	3	3
Zinc, yellow passivate	1	1	1	1	3	1
Zinc, olive passivate	1	1	1	1	2	1
Tin, no passivate	1	3	1	3	1	3

Grading : 1   little or no attack
2   slight attack
3   substantial attack

Table 4 Effect of zinc passivate treatment in neutral salt spray

Passivation and drying treatment for 3 μm zinc	48 Hours		96 Hours		240 Hours	
	White Rust	Red Rust	White Rust	Red Rust	White Rust	Red Rust
iridescent	1	1	1	1	1	1
olive green	1	1	1	1	2	1
black	1	1	1	1	2	1
iridescent + hot air drying	1	1	1	1	3	1
olive green + hot air drying	2	1	2	1	3	1
black + hot air drying	2	1	2	1	3	1

Table 5 Accelerated humidity testing[*]

Coating system (3 μm metal)	White rust	Red rust
Cadmium unpassivated	1	2
Cadmium yellow passivate	1	1
Zinc unpassivated	3	1
Zinc yellow passivate	1	1
Zinc olive passivate	3	1
Zinc black passivate	2	1
Tin unpassivated	1	3

* Six 24 hour cycles of 16 hrs. at 55°C and 8 hrs. at 25°C, humidity being maintained at 95% min.

Table 6 Neutral salt spray test data[*]

Coating type	Time in Hours to red rust for various thicknesses (μm)				
	5	6.5	7.5	10	13
Plain cadmium	50		120	200	320
Cadmium + chromate	200-500	200-500	500-2000	500-2000	
Cadmium-tin (50%)	50		120	200	320
Cadmium-tin (50%) + chromate	200-500	200-500	500-2000	500-2000	
Plain zinc	36		48		96
Zinc + heavy chromate	100-250	100-300	120-400	150-400	150-500
Zinc-nickel alloy (11% Ni)		80-100	100-200	200-350	500-1000
Zinc-cobalt alloy (0.8% Co)		100-200	150-250		
Passivated zinc-cobalt alloy (0.8% Co)		900-1200			
Zinc-iron alloy					
Zinc-manganese alloy (50% Mn)			800-1200		
Zinc-tin alloy (80% Sn)			600-800		

* Data obtained from many (not always comparable) sources

are different the environments may be similar and parallels in corrosion performance can reasonably be drawn.

6 CONCLUSIONS

A number of alternatives to cadmium exist and provided a careful choice is made equivalent performance can be achieved.

A tin-based coating would only be used if solderability is a vital product requirement because the corrosion performance is generally inferior.

Zinc and zinc alloy coatings provide the best corrosion performance and if an appropriate thickness and passivation treatment are specified corrosion performance may actually be enhanced.

REFERENCES

1. K. Nomiyama. Fifth International Cadmium Conference, San Fancisco, 1986.
2. E.L. Anderson. ibid.
3. D.R. Gabe. Bull. Inst. Corr. Sci. Tech. 1981, _19/31_, 2.
4. D.N. Layton. Trans. IMF 1965, _43_, 153.
5. H. Simon. Metalloberflache 1982, _36_, 1.
6. F. Albers. Galvanotechn. 1982, _73_, 736.
7. J.S. Hadley. IMF Symposium 'Environmentally Friendly Surface Finishing', 1990.
8. Anon. Metal Progress 1972, 74-77.
9. C. Hoskins. Production 1974, (1), 89-91.

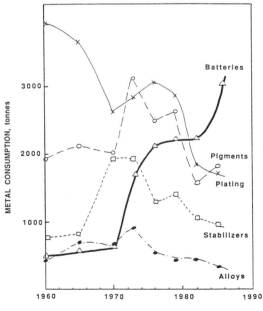

Figure 1

Usage of cadmium for the

period 1960-1985

1.5.2
Electrodeposited Coatings Containing Alumina in a Nickel–Phosphorus Matrix

D. B. Lewis, G. W. Marshall, and B. E. Dodds

MATERIALS RESEARCH INSTITUTE, SHEFFIELD HALLAM UNIVERSITY, SHEFFIELD S1 1WB, UK

1   INTRODUCTION

The electroless deposition of Ni-P coatings is well established.[1]   The formation and properties of the coatings containing dispersions of alumina and carborundum have been studied,[2,3] and such coatings have been produced commercially by electroless deposition.

Coatings containing dispersions of ceramic particles in simple metal matrices have also been studied extensively.[4,5,6,7,8]   Broesziet et al reviewed the properties of electrodeposited nickel coatings dispersion hardened with alumina and silicon carbide.[7] Recently Celis et al reviewed the theories relating to the formation of electrodeposited metal-ceramic coatings with particular respect to copper-alumina deposits.[8]

Electrodeposited Ni-P coatings have been prepared by conventional plating and their properties studied.[9,10,11]   Other workers have investigated the use of pulse plating[12] and superimposed pulsed current electrolysis on electroless plating baths to produce Ni-P coatings.[13,14,15]

Although the use of electroless processes is widespread in the commercial production of Ni-P coatings, their operation has a number of drawbacks. Therefore the development of electrodeposition processes for the production of Ni-P coatings containing dispersed ceramic particles might prove attractive.   Such processes would offer the prospect of tailoring the deposit's physical properties through a combination of the control of the coating composition coupled with the use of suitable heat treatments on the 'as plated' deposits.

2   EXPERIMENTAL PROCEDURE

Preparation of Electrodeposited Coatings

Prepared copper sheet samples were weighed and plated using either conventional direct current or direct pulsed current. The latter was operated with an on to off ratio of 10:100 ms. The pulse current densities quoted refer to the average value taken over the full plating cycle.

The composition of the plating bath used was that developed by Brenner et al and later used by Lashmore (see Table 1).

The solution had a pH of 0.8 and was operated at 80 °C using two nickel anodes, one placed either side of the cathode, immersed in 2l of the bath. Bath agitation and suspension of alumina where appropriate was achieved using a combination of a propeller stirrer and a submersible non-vortex magnetic stirrer.

Plating was carried out at current densities of 10, 40, 60 and 80 mA cm^{-2} in baths operated with and without the addition of 15 g l^{-1} α-alumina. The alumina used was 1 μm polishing grade (Banner Scientific Ltd, Coventry). The same quantity of current was passed during each plating run.

Post Plate Treatment and Examination of Coatings

Heat Treatment. All heat treatment of plated specimens was at 400 °C for 60 minutes in an inert atmosphere.
Metallurgical and SEM Examination. Metallurgical and SEM examinations were carried out on specimens having a coating thickness of at least 20 μm.

Structural studies by X-ray diffraction using monochromatic CuK$_\alpha$ radiation were carried out directly on specimens in either the 'as plated' or heat treated condition. The specimens were analysed by X-ray fluorescence for the elements nickel, phosphorus and aluminium. The weight percent alumina in the deposit was then determined by assuming stoichiometry. Where appropriate, the specimens were then prepared for further examination. Thus specimens for SEM examination and hardness measurements were overplated with a minimum of 50 μm copper. These were then mounted in conductive Bakelite and prepared in the usual manner to a one

Table 1   The Composition of the Ni-P Plating Bath

Nickel Sulphate	($NiSO_4.6H_2O$)	150 g/l
Nickel Chloride	($NiCl_2.6H_2O$)	50 g/l
Phosphoric Acid	(H_3PO_4)	50 g/l
Phosphorous Acid	(H_3PO_3)	40 g/l

micron finish.

Specimens containing alumina were examined using scanning electron microscopy to assess the distribution of alumina within the coatings using secondary and backscattered electron image techniques. Specimens were also analysed for nickel and phosphorus by energy dispersive analysis (EDX) at points in the Ni-P matrix. The analytical results were quantified using the standard ZAF technique. Analyses of the coating were carried out, both adjacent to the copper substrates and near to their surfaces to assess the uniformity of the composition within the coatings.

Microhardness Measurements. Microhardness measurements were made using a 25 g load and the final values obtained are based upon the average of eight indentations.

3   EXPERIMENTAL RESULTS

Plating Efficiencies

The overall plating efficiencies were calculated for the deposition process based upon the masses of nickel and phosphorus deposited during conventional and pulsed current plating runs. It was assumed that the mechanism of alumina inclusion involved a mechanical, rather than a direct electrolytic process and therefore used no current directly.

It was found that the overall current efficiency decreased with increasing current densities in all cases (see Figure 1). The current efficiencies were also lower for baths containing alumina suspensions than when they were operated at the same current density without alumina. This was true for both conventional and pulsed current.

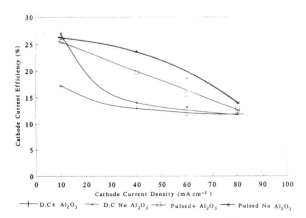

Figure 1 Variation of cathode current efficiency for Ni-P Deposition as a function of current density

Figure 1 also shows that the current efficiencies were greater for baths operated using pulsed rather than conventional current.

Structural Studies

As-deposited Coatings. Figure 2 shows typical X-ray diffraction traces obtained for conventional and pulse plated deposits containing no alumina. It is clear that the deposits contained amorphous nickel. However, some deposits also contained a crystalline phase (see trace 'b' in Figure 2) which was identified as Ni_5P_2. The presence of this phase was more prevalent in deposits formed using higher current densities.

As-deposited Coatings containing Alumina. The Ni-P matrix structure in the deposits containing alumina and plated using both conventional and pulsed current mirrored those without alumina. Alumina was only detectable in coatings produced using lower current densities, eg 10 mA cm^{-2} (see Figure 2c).

Heat Treated Deposits. Typical X-ray diffraction traces obtained from conventional and pulse-plated deposits with and without alumina are included in Figure 3. The structures of the heat-treated deposits are dealt with in terms of the phases present in the Ni-P matrix. Figure 3 relates to the structure of the Ni-P matrix for a deposit obtained at 10 mA cm^{-2} and contains mainly Ni_3P with small amounts of nickel and Ni_5P_2. The presence of alumina or the use of pulsed current had little effect on the phases formed within the matrix. The CuK_α doublets of the Ni_3P lines were partially resolvable, thus indicating negligible line broadening, in the deposit obtained at 10 mA cm^{-2} and following heat treatment. This indicated a stress free deposit with a grain size of at least 0.5 μm.

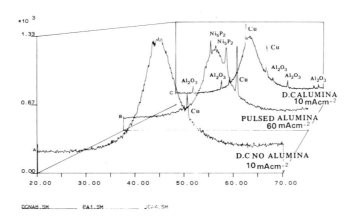

<u>Figure 2</u> X-ray diffraction traces of the as-deposited coating

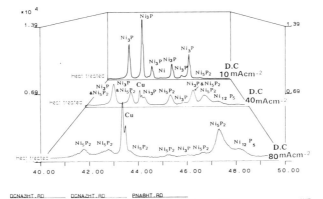

<u>Figure 3</u> X-ray diffraction traces of heat treated coatings showing the effect of current density

X-ray diffraction traces obtained after the annealing of conventionally plated deposits using 40 and 80 mA cm^{-2} are shown in Figure 3. The structure of the former consisted of a mixture of Ni_3P, Ni_5P_2 and minor amounts of $Ni_{12}P_5$. The apparent volume fractions of the Ni_5P_2 and $Ni_{12}P_5$ phases increased, with a corresponding decrease of the Ni_3P phase, for deposits obtained at successively higher current densities. Ni_5P_2 appeared to be the major phase present in deposits produced at 80 mA cm^{-2}.

X-ray diffraction studies of specimens with alumina and plated conventionally using 80 mA cm^{-2} are shown in Figure 4 which indicates the presence of Ni_3P, Ni_5P_2, $Ni_{2.55}P$ and $Ni_{12}P_5$ phases. The trace from the conventionally plated deposit without alumina is also included in Figure 4 for comparison. However, the apparent volume fraction of Ni_3P was greater than that of the corresponding deposit formed without alumina.

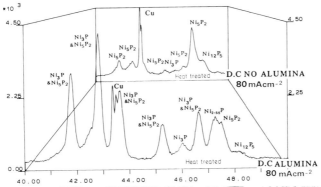

<u>Figure 4</u> X-ray diffraction traces of heat treated coatings at 80 mA cm^{-2} for DC plated with and without alumina

Metallography. A series of backscattered electron images showing the distribution of alumina in conventionally plated deposits is included in Figure 5. The letters A, B and C on these micrographs refer to the overplate, deposit and substrate respectively. For both types of deposit, the volume fraction of alumina was the greatest in deposits produced at the lowest current density, 10 mA cm^{-2}. The volume fraction of alumina decreased as the current density used to form the deposit increased. The alumina was distributed non-uniformly in all the deposits and a high degree of agglomeration of the particles was evident.

Deposit Composition. The compositions of the deposits are given in Table 2, which also gives the Ni-P matrix composition for Ni-P-Al$_2$O$_3$ coatings.

It can be seen that, in general, the phosphorus content of the coatings increased as the current density used to form them increased whilst in contrast the alumina content decreased. The phosphorus content of conventionally plated coatings tended to be greater than the corresponding coatings obtained by pulse plating. In contrast, the reverse tendency is observed with respect to their alumina content.

Microhardness. The microhardness measurements made on conventionally and pulse-plated deposits with and without alumina dispersions are given in Table 3. Hardness values are given for deposits in both the as-plated and heat-treated conditions.

Conventionally as-plated coatings containing dispersions of alumina were consistently harder than corresponding coatings containing no alumina. However, following heat treatment the coatings with and without alumina dispersions had similar hardness which showed an almost twofold increase over their 'as-plated' values.

D.C. PLATED SAMPLES

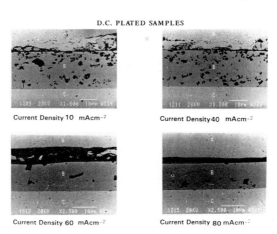

Current Density 10 mAcm-2   Current Density 40 mAcm-2

Current Density 60 mAcm-2   Current Density 80 mAcm-2

Figure 5 Scanning electron micrographs showing the distribution of alumina within the coating

Table 2 The Percentage Composition of Electrodeposited Ni-P and Ni-P-Al$_2$O$_3$ Deposits

Current Density mA cm^{-2}	NiP Ni - P		Ni - P - Al$_2$O$_3$ Ni - P - Al$_2$O$_3$		NiP Matrix only Ni	P
10 DC	84.8	15.2	77.48 13.67	8.8	85	15
40 DC	83.9	16.1	76.78 14.6	8.6	84	16
60 DC	82.9	17.1	80.83 15.3	3.7	84	16
80 DC	83.0	17.0	80.85 15.7	0.4	84.2	15.8
10 P	84.6	15.4	80.47 13.7	5.76	85.4	14.6
40 P	84.5	15.5	80.35 15.8	3.77	83.5	16.5
60 P	84.2	15.8	83.2 15.87	0.81	83.9	16.1
80 P	84.0	16.0	83.41 15.89	0.69	84.0	16.0

DC = Conventional Plated   P = Pulsed Plated

Pulse-plated coatings with and without alumina were slightly harder than any of the 'as-plated' coatings produced by conventional plating. With heat treatment the hardness of pulse-plated deposits with and without alumina dispersions increased. However, the hardness of coatings with and without alumina were the same following heat treatment.

Table 3 Microhardness of Electrodeposited Ni-P and Ni-P-Al$_2$O$_3$ Deposits, with and without Heat Treatment

Current Density mA cm^{-2}	Microhardness Hv 25			
	Without Alumina		With Alumina	
	As-Deposited	Heat-Treated	As-Deposited	Heat-Treated
10 DC	449	920	518	901
40 DC	402	902	498	870
60 DC	392	900	510	901
80 DC	405	893	491	897
10 P	543	891	527	913
40 P	515	897	538	878
60 P	530	879	545	897
80 P	522	886	524	903

DC = Conventional Plated   P = Pulsed Plated

4   DISCUSSION

The Structure of Deposits

As-Deposited Coating. X-ray diffraction studies show that in some cases the matrix of conventionally and pulse-plated Ni-P and Ni-P-Al$_2$O$_3$ deposits are amorphous. This finding is in keeping with earlier work [15,16,17,18] which suggested that coatings containing more than 7 mass percent phosphorus have an amorphous structure. However, some of the deposits also contained a crystalline phase which was identified as Ni$_5$P$_2$. The presence of this phase was not particularly associated with either conventional or pulse-plating. However, it was prevalent at higher current densities and hence higher phosphorus contents. The presence of the two amorphous phases corresponding to Ni$_3$P and Ni$_5$P$_2$ have been reported.[12] However, it is not clear why the Ni$_5$P$_2$ crystalline phase is present within the coatings.

Heat Treated Deposits. Conventional and pulse-plated Ni-P and Ni-P-Al$_2$O$_3$ (Ni-P matrix) change from amorphous to partially crystalline as a result of annealing. The matrix composition of the coatings ranges from 14.6 to 17.1 wt % phosphorus.

The equilibrium diagram,[19] predicts that the lower phosphorus content lies just within the Ni$_3$P-nickel phase field, whereas the higher phosphorus content lies just within the Ni$_3$P-Ni$_5$P$_2$ phase field. The stoichiometric compositions of nickel phosphide phases identified in this investigation are given in Table 4.

The structure of the nickel-phosphorus matrix in all cases consisted mainly of Ni$_3$P with minor amounts of nickel and Ni$_5$P$_2$ even though the compositions carried between 14.6 and 15.4 wt % in deposits obtained at 10 mA cm^{-2}. However, with the exception of the coating formed by pulse plating at 10 mA cm^{-2}, all the compositions lie within the Ni$_3$P-Ni$_5$P$_2$ phase field. Hence only phases Ni$_3$P and Ni$_5$P$_2$ would be predicted from the phase diagram for the other coatings plated at a current density of 10 mA cm^{-2}. The reason for these unpredicted phases from the equilibrium diagram is thought to result from constituent segregation.[20] Since all compositions lie close to the Ni$_3$P phase boundary, localised variations in phosphorus content within the coating results in the presence of Ni$_3$P, Ni$_5$P$_2$ and nickel in all deposits plated at a current density of 10 mA cm^{-2}. Ni$_3$P is the major phase since the compositions of all deposits lie close to the Ni$_3$P phase boundary. The Ni$_5$P$_2$ increases at the expense of the Ni$_3$P for coatings with alumina obtained with increasing current densities. This is predictable from the equilibrium[19] diagram since coatings obtained at 40 and 80 mA cm^{-2} contain 16.1 and 17.0 wt % phosphorus respectively.

Table 4   The Stoichiometric Composition of Nickel-
Phosphide Phases

Nickel Phosphide Phase   Composition (Wt % P)

Ni_3P 14.9
$Ni_{2.55}P$ 17.14
Ni_5P_2 17.4
$Ni_{12}P_5$ 18.0

However, at both 40 and 80 mA cm^{-2} the presence of
a nickel phosphide phase $Ni_{12}P_5$, not reported in the
equilibrium diagram, was also identified and its
apparent volume fraction increased with increasing
phosphorus content.   The presence of $Ni_{12}P_5$ has been
reported in Ni-P alloys with phosphorus contents greater
than 17.4 mass % [21,22].   The reason for the presence of
phases not predicted from the equilibrium diagram,
$Ni_{12}P_5$, is thought to result from constituent
segregation.   Localised segregation of phosphorus leads
to regions with compositions outside the Ni_3P - Ni_5P_2
phase field resulting in the formation of $Ni_{12}P_5$.   It
can be postulated that, due to the higher phosphorus
content, coatings plated at a current density of 80 mA
cm^{-2} (17 wt % P compared to 16.1 Wt % P) have a greater
apparent volume fraction of $Ni_{12}P_5$.

In the coating plated at a current density of 80 mA
cm^{-2} with alumina, the apparent volume fraction of Ni_3P
was greater than that in the coating without alumina.
This was because of the lower phosphorus content in the
coating containing alumina (15.8 wt % P compared with 17
wt % P).   The presence of $Ni_{2.55}P$ has been reported,[10]
in nickel phosphorus alloys and often forms in
preference to or together with Ni_5P_2.

Hence, the type and apparent volume fraction of
phases present in these alloys is a function of their
phosphorus contents.   The phases present and the
response to heat treatment of the electroplated deposits
(using conventional and pulsed current) were similar to
those produced by the more conventional electroless
method.
The Plating Process.   The decrease of the plating
efficiency with increasing current density observed for
the bath operated with both conventional and pulsed
current is in keeping with the behaviour of many acid
plating baths.   Similarly, the increase in efficiency
found when using pulsed rather than conventional DC
current is well-established.   However, the apparent
effect of alumina suspensions within the plating bath,
operated in either the conventional or pulsed current
mode, to reduce its efficiency appears anomalous.
Normally increased agitation reduces concentration over
potentials and hence tends to increase cathode current
efficiency especially at higher current densities.   An
explanation probably lies in the abrasive nature of the

aluminium particles which probably cause simultaneous wear of the growing deposit due to their scouring action. Thus the efficiencies recorded are probably not the true electrochemical ones.

Clearly the work has established that $Ni-P-Al_2O_3$ coatings can be produced by electrodeposition as well as electroless deposition. However, the combination of the low current density used to produce the coatings (10 mA cm^{-2}) coupled with the low current efficiency of the process make it impractical from a commercial point of view. Further work is therefore being carried out to develop a version of the bath which will allow a greater plating rate. At present, baths containing lower concentrations of phosphorous acid are being studied with a view to producing $Ni-P-Al_2O_3$ coatings at a rate in keeping with commercial practice.

REFERENCES

1. W. Riedel, 'Funktionelle Chemische Vernicklung', Eugen. G. Leuze, Verlag, Saulgau/Württ, 1989.
2. W. Metzger and R. Ott, Metalloberfläche, 1979, 33, 456.
3. W. Metzger and Th. Florian, Metallkunde, Proc. DGfMK e.V. (1979), 71.
4. W. Metzger and Th. Florian, Metalloberfläche, 1980, 34, 274.
5. F. Sautter, Metall, 1964, 18, 596.
6. V. P. Greco and W. Baldauf, Plating, 1968, 3, 250.
7. E. Broszeit, G. Heinke and H. Weigand, Metall, 1971, 25, (5), 470.
8. J. P. Celis, J. R. Roos, C. Buelens and J. Fransaer, Trans. Inst. Metal Finish., 1991, 69, 133.
9. A. Brenner, 'Electrodeposition of Alloys', Academic Press, New York, 1963.
10. U. Pittemann and S. Ripper, Z. Metallkunde, 1983, 74, 783.
11. A. Brenner, D. E. Conch and E. K. Williams, J. Research NBS, 1950, 44, 109.
12. D. S. Lashmore and J. Weinroth, Second International Pulse Plating Symposium, 6-7 October 1981, Rosemont, Pub. Am. Electrodep. Soc. Inc., Florida.
13. H. Koretzky, U.S. Patent 3,485,725, 23rd December 1969.
14. P. Cavallotti, L. Nobile, D. Colombo and F. Kruger, Trans. Inst. Metal Finishing, 1990, 28, 55.
15. G. W. Marshall, D. B. Lewis and B. L. Dodds, to be published.
16. H. Kreye, H. H. Müller and T. Petzel, Galvanotechnik, 1986, 77, 561.
17. S. V. S. Tyagi, Z. Metallkunde, 1985, 76, 492.
18. Q. X. Mai, R. D. Daniels and H. B. Harpalani, Thin Solid Films, 1988, 166, 235.
19. M. Hansen, 'Constitution of Binary Alloys', McGraw-Hill, New York, 1965, 1027.

20. M. A. Erung, L. Shoufu and L. Pengxing, <u>Thin Solid Films</u>, 1988, <u>166</u>, 273.
21. S. Rundqvist and E. Larson, <u>Acta Chem. Scand.</u>, 1959, <u>13</u>, 551.
22. S. Rundqvist, <u>Acta Chem. Scand.</u>, 1962, <u>6</u>, 992.

1.5.3
Laser Remelting of Flame-sprayed Coatings

K. Wang, Y. Zhu, W. Wei, Z. Tian, and C. Song

DEPARTMENT OF MECHANICAL ENGINEERING, TSINGHUA
UNIVERSITY, BEIJING 100084, PEOPLE'S REPUBLIC OF CHINA

1 INTRODUCTION

Because of its good corrosion resistance, austenitic
stainless steel has wide applications in industry, but its
wear-resistance is poor. It has a large coefficient of
friction and it is susceptible to adhesion. It can not
meet the demand for materials which should have not only
good corrosion resistance but also good wear resistance.
Therefore the potential of raising the wear resistance of
stainless steel is attractive.

The coating method is a widely used technology to
improve the performance of a material's surface. Ni-based
alloys having not only good corrosion resistance but also
good wear resistance, are candidate materials for coating
stainless steel.

In the present work, the flame-sprayed method has
been used to spray Ni-based alloys on the surface of
stainless steel. Assessment indicates that the coatings
have good wear resistance. Nevertheless flame-sprayed
coatings have characteristic problems of for instance,
porosity, interparticle cohesion and low bonding strength
between substrate and coatings[1,2]. These attributes
significantly influence the coating system's ability to
resist wear and corrosion.

Laser surface treatment has been employed to remelt
and recrystallize flame-sprayed coatings. Good results
have been obtained. The microstructure of laser remelted
coatings is extremely fine, pore-free and metallurgically
bonded to the substrate. The hard particle phases, such

as CrB, Cr_2B, W_2C and Ni_3Si, exist in the coatings after laser remelting.

The wear test shows that the wearing capacity (material removal rate) of the coatings is very small. These coatings can confidently be expected to meet the demand for good wear and corrosion resistance.

2 EXPERIMENTAL METHODS

The experimental substrate material was 1Cr18Ni9Ti austenitic stainless steel. The dimensions of the samples were 70x30x4 mm.

The alloy powders sprayed were Ni-based. The chemical composition of the alloys is shown in Table 1. The three alloy powders have good wear performance and corrosion resistance. Their hardnesses are as follows: Metco16C is HRc60, Metco31C is HRc62 and Ni45B is HRc45.

The samples were cleaned, sand blasted and preheated before flame-spraying. The thickness of the sprayed coatings was 0.3-0.4 mm.

A 2 kW continuous wave CO_2 laser was used to remelt the flame-sprayed coatings. The laser powers used were 800, 1000 and 1400 W. The diameter of the spot was 3 mm. The scanning speeds were 150, 300 and 450 mm/min. The scanning overlap area was 50%.

The usual methods of analysis were employed; metallography and SEM for microstructure, EDX for elemental composition and X-ray diffraction for phase constitution. Microhardness and wear of the coatings systems were characterized using a HX-200 microhardness tester and MHK-500 ring block tester.

3 EXPERIMENTAL RESULTS AND ANALYSES

Microstructure of the Coatings Before and After Laser Remelting

The microstructures, before laser remelting, of the three coatings have some common characteristics. The microstructure of the Ni45B coating is shown in Figure 1. It can be seen that the sprayed coating is a loosely

packed grainy layered structure. The bonding within the
coating itself and between the coating and substrate is
poor. The original alloy powders mainly took the form of
inlayed, squat particles except for a small amount of
sphericity. There are small openings and impurities
visible in the coating, and a gap between the coating and
substrate. Mechanical bonding is apparent.

The structural characteristics of a sprayed coating
are determined by the defining parameters of the flame-
spray. During the flame-spray process, the material
sprayed is heated until molten, half-molten or in a highly
plastic state. It rapidly penetrates the substrate and
deforms into a flaky grain. The later arriving material
impacts the grains on the substrate and also produces
flaky grains, which inlay each other and a coating is
gradually formed. The coating is directional in nature
and has some porosity. In general, the coating and
substrate can only form mechanical bonding, even though
the surface of the substrate was sand blasted. At best
when particles sprayed penetrate the substrate, micro-
diffusion may

Table 1 Chemical composition of alloy powders (wt.%)

Powder	C	Cr	B	Si	Fe	Mo	Cu	Ni	Co-WC
Metco16C	0.5	16	4	4	2.5	3	3	(67)	–
Metco31C	0.5	11	2.5	2.5	2.5	–	–	46	35
Ni45B	0.3-0.6	10-11	2-3	3-4.5	<17	–	–	Bal.	–

occur at the interface through close proximity of the
particles and the substrate at high temperature. There
may be a bonding force between the particles and substrate, but it is very difficult to form a metallurgical
bond[3]. The property of the coating can be improved through
optimization of spray parameters, but it is impossible to
eliminate completely coating defects which may affect the
use of the coating in the field of wear and corrosion.

After laser remelting, the sprayed coating is level and smooth, however its microstructure changes. The microstructure of the surface and intersection after laser remelting is shown in Figure 2. It can be seen that there are several characteristics in the coating microstructure. Firstly, the microstructure of the remelted coating is changed to fine-grained. When a laser beam scans the surface of the sample, the coating and surface of the substrate are heated rapidly to the molten state and cooled quickly after the laser pass. There are many foreign particles in the molten pool because of the alloy elements added. These foreign particles are crystal cores which have become solid before the grains grow large. Therefore more homogeneous and finer crystalline grains are formed[4]. It has been observed that the grains become more homogeneous and fine with an increase in scanning speed with fixed laser power or with a decrease of laser power with fixed scanning speed. The reason is that the cooling speed increases in these conditions. Secondly, the morphology of the coating consists mainly of tree-like crystals and equiaxed crystals. The morphology at the interface between the coating and the substrate is mainly fine tree-like crystals and these dendritic crystals tend to grow vertically to the interface. This morphology is produced by the high temperature gradient existing between the molten area and the substrate, grains grow along the direction of heat diffusion. Thirdly, the bonding within the coating itself and between coating and substrate is improved. When the coating is molten and recrystallized, the impurities, such as oxides, float easily in the liquid state and move outside with laser scanning. The porosity is basically eliminated. The microstructure of the coating becomes homogeneous, fine and dense. The toothed mechanical keying between coating and substrate is transformed to a smooth, level metallurgical bond.

Chemical Composition and Structure of the Coating

After laser remelting, the chemical composition of the coating changes. The altered chemical composition of three coatings is shown in Table 2. The parameters for the laser remelting are as follows: power 1000W and scanning speed 300 mm/min. It can be seen that Fe and Cr increase and Ni decreases after laser remelting. The main reason is that the coating and part of the substrate are melted during laser remelting, their elements mix in the

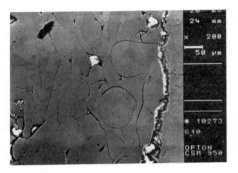

<u>Figure 1</u> Microstructure of Ni45B coating before remelting

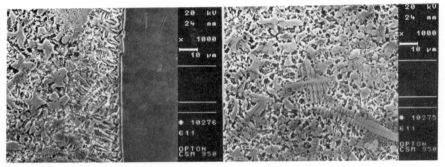

<u>Figure 2</u> Microstructure of coating after remelting

molten pool and the coating is diluted by the elements of
the substrate, such as Fe. Furthermore the diffusion of
Fe and alloy elements can also affect the chemical
composition of the coating.

A model describing the interfacial reaction is needed to
explain the diffusion phenomenon. A certain number of
atoms, such as Fe, at the surface of the substrate can
receive sufficient energy and be in a heat-active state
because of induction of the alloy liquid. These atoms can
separate from the substrate lattice and move into the
coating - that is physical dissolution. At the same time
the alloy elements in the molten pool can dissolve the r-phase
at high temperature through solid diffusion - i.e. r-phase
alloying. The variation in chemical composition of the
coating is beneficial in forming a metallurgical coupling
between the coating and the substrate.

The structure of the coating was analysed by an X-ray
diffractometer. The results were as follows: there were
r-(Ni,Fe), Cr_3C_2 and CrB and there may be Ni_3Si, Ni_3B and

<u>Table 2</u> Chemical composition of coatings (wt.%) after
laser remelting

Powder	C	Cr	B	Si	Fe	Mo	Cu	Ni	Co–WC
Metco16C	–	20.4	–	4.57	6.42	–	2.54	62.66	–
Metco31C	–	17.3	–	8.14	7.05	–	–	36.51	Co3.05 W26.64
Ni45B	–	16.2	–	3.19	21.37	–	–	59.22	–

$Fe_{23}(C,B)_6$ in the Metco16C coating; there were r-(Ni,Fe),
Cr_3C_2, Cr_2B, W_2C, Ni_3Si_2, $(Fe,Ni)_{23}C_6$ and $Ni_{31}Si_{12}$ and perhaps
also Ni_3B and $M_{23}C$ in the Metco31C coating.

When a laser beam was scanning the surface of the
sample, additive alloy elements and the substrate surface
were in the liquid state. The alloy elements and
substrate elements were mixed and combined into chemical
compounds based on B, C, and Cr; these were hard particles
and were distributed over the whole surface. Therefore a
wear resistant coating was formed at the substrate
surface.

<u>Microhardness of the Coating Before and After laser
Remelting</u>

The hardnesses of the three flame-sprayed coatings
before laser remelting were as follows: Metco31C was
$HV_{0.1}773$, Metco16C was $HV_{0.1}667$ and Ni45B was $HV_{0.1}557$. The
hardnesses of the coatings after laser remelting are shown
in Table 3.

The hardness of the coatings after laser remelting
decreased compared with before laser remelting. This was
related to the dilution of the substrate to the coating.
The defining parameters of the laser have some influence
over the hardness of the coating. When the laser power
was fixed, the hardness increased with an increase in
laser scanning speed. When the scanning speed was fixed,
the hardness decreased with an increase in laser power.

<u>Table 3</u> Microhardness of coatings after laser remelting ($HV_{0.1}$)

Powder	Power, W	Scanning Speed, mm/min.		
		150	300	450
Metco16C	1000	342	440	512
	1400	306	360	478
Metco31C	1000	358	475	558
	1400	342	405	508
Ni45B	1000	310	362	473
	1400	220	311	445

When the laser power was fixed, the higher the scanning speed, the less the substrate melted and the dilution of the coating decreased. The temperature of the substrate increased less and the cooling rate of the coating was faster, and formed a fine and homogeneous microstructure and increased the hardness of the coating. When the scanning speed was fixed, the higher the power, the more the substrate melted; thus the dilution of the coating increased, the temperature of the substrate increased and the cooling speed of the coating was reduced. Therefore the hardness decreased.

<u>Wear Resistance of the Coating After Laser Remelting</u>

Using a MHK-500 ring-block wear tester the friction force and friction factor can be determined dynamically. The sample block was rubbed against a GCr15 ring. After a certain time, a Talysurf 5p-120 measurement system was used to record the appearance of the transections of the worn coating. Then the wear area of transection and wearing capacity were calculated. The wear test used a light load, slow speed and short time to suit the characteristics of the coating and the substrate. The experimental parameters were: load 4.0 kg, running speed 300 rpm, 6 minutes duration and No.30 engine oil.

When the laser power was 1000W and the scanning speed of the laser was 300 mm/min., the friction coefficients of the three coatings were: Metco16C 0.1523, Metco31C 0.1109 and Ni45B 0.1052.

The stainless steel substrate was rather soft, with an $HV_{0.1}$ of 108. When it was rubbed against other materials, adhesive wear often occurred and hence the friction coefficient was rather large (0.35-0.67). However, the friction coefficient of the coating was much lower than that of stainless steel.

Table 4 shows the wearing capacities calculated for the three coatings. It can be seen that the wearing capacities of the three coatings were small, which indicates that they all have good wear resistance. The wearing capacities of the three coatings were all different and reflected different wear resistance, with Metco31C having the best wear resistance.

Table 4 Wearing capacity

Powder	Wear Parameters	
	Transection Area mm^2	Wearing Capacity mm^3
Metco16C	1.7×10^{-3}	0.02125
Metco31C	5×10^{-4}	0.005625
Ni45B	9×10^{-3}	0.1125

4 CONCLUSIONS

The properties of the coating were determined by, inter alia, chemical composition and microstructure. These three coatings were composed of Ni, Cr, B and Si alloy elements. These elements and elements from the substrate, such as Fe, combined to give particles possessing high hardness, which were beneficial in raising wear resistance. Furthermore:

1. Laser remelting improves bonding within the coating and between the coating and the substrate, and achieves a fine microstructure, free from porosity and impurities, thus enhancing the quality of the coating.
2. The coating was diluted by the substrate after laser remelting. The chemical composition of the coating changed. The structure of the coating was mainly

chemical compounds of Cr, Ni, B and C, which possess high hardness.

3.   The hardness of the coatings decreased after laser remelting.

4.   The friction coefficients and wearing capacities of the laser remelted coatings were small.

REFERENCES

1.   C. Shen, 'Flame-Spray and Spray-Welding', Railway Press of China, Beijing, 1984.

2.   F. Wei, 'Technology of Thermal-Spray', Mechanical Industry Press.

3.   M. Xiang, Master's Thesis, Tsinghua University, 1987.

4.   K. Zheng and S. Zhang, <u>Laser Tech.</u>, 1991, <u>15</u>.

1.5.4
The Corrosion Behaviour of Molybdenum Plasma Sprayed Coatings on Steel

A. Koutsomichalis, H. Badekas, and S. Economou

LABORATORY OF PHYSICAL METALLURGY, NATIONAL TECHNICAL
UNIVERSITY OF ATHENS, GREECE

1 INTRODUCTION

Plasma spraying is a versatile technique for applying
protective coatings to engineering alloys. Plasma sprayed
coatings are often used for thermal oxidation, corrosion
and wear protection of various substrates[1-3]. These
coatings are frequently exposed in corrosive environments
even when their major purpose is not to provide corrosion
protection.

Plasma sprayed coatings exhibit a degree of porosity
because the molten droplets deposited by the plasma
spraying process cool very rapidly and cannot completely
wet the substrate surface. This residual porosity affects
the corrosion behaviour of plasma sprayed coatings because
corrosive fluids can penetrate the coating and dissolve
the substrate material[4]. Furthermore, galvanic effects
between the coating and the substrate must be taken into
account[5].

Among the several plasma sprayed metallic coatings
which are used today are the molybdenum coatings; these
coatings have excellent wear resistance, particularly in
erosion-cavitation wear but are sensitive to air oxidation
at temperatures exceeding 650^0C.

The main purpose of the present work was to examine
the morphology, the composition and the corrosion
behaviour of molybdenum plasma sprayed coatings on mild
steel specimens.

2 EXPERIMENTAL

Pure molybdenum powder (99.5 wt.%) having a size
distribution in the range of -99 + 44 µm was plasma
sprayed on the surface of mild steel specimens. The

substrate was a SAE 4130 steel with the following
composition: 0.28-0.33 wt.% C, 0.95 wt.% Cr, 0.20 wt.% Mo
and 98.5 wt.% Fe. Each of the steel specimens was grit
blasted prior to plasma spraying.

The plasma spraying equipment used was a METCO 7MB
unit and the spray parameters are listed in Table 1. The
spraying was always performed in air and the distance
between the spraying gun and the substrate was 13 cm. The
coating thicknesses were 200, 250, 325 and 450 μm. A cold
air stream was blown on the back surface of each steel
specimen during spraying thus keeping the temperature of
the substrate below 150^{0}C.

<u>Table 1</u> Plasma spray parameters

1. GAS PARAMETERS		
	Pressure (MPa)	Flow rate (l /min)
Primary gas (N_2) Secondary gas (H_2) Carrier gas	0.379 0.379 0.379	495.6 71.0 170.0
2. GUN PARAMETERS		
Current: 400 A Voltage: 80 V Spray rate: 4.6 Kg /hr		

The morphology of the molybdenum coating was studied
with the help of an optical microscope and a Jeol 35 CF
scanning electron microscope. The corrosion behaviour of
the coated and uncoated specimens was examined in an HCl
solution (pH=1.5) at 20^{0}C by measuring the potential of the
immersed specimen in the solution vs. standard calomel
electrode as a function of immersion time. Samples of the
corrosive solutions were analysed with the aid of a
PERKIN-ELMER 3000 atomic absorption instrument.

The data presented in this study are the mean values
of five independent experiments.

3 RESULTS AND DISCUSSION

Figure 1 shows the transverse section of the molybdenum coating deposited on the surface of the steel specimen. Microstructural inhomogeneity within this plasma sprayed coating is evident in this micrograph. The coating has a laminated structure because of the flattening that the sprayed particles undergo upon impacting on the substrate. The coating - substrate interface is shown to be roughened and deformed due to the grit blasting; the serrations which are present in the coating follow the serrations at the interface.

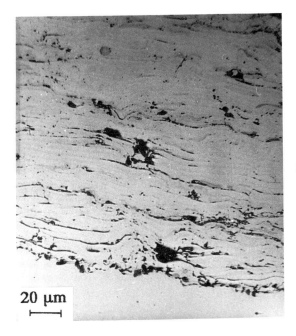

molybdenum

20 µm

steel

<u>Figure 1</u> Transverse section of the molybdenum plasma sprayed coating deposited on steel

Porosity is typically present in this microstructure and is caused by air entrapment among the impacting particles. The mean porosity of the molybdenum coatings was measured (using image analysis) to be equal to 1.1%. Coarse pores were observed having a mean size of 6-8 µm as well as fine pores with an average size of 0.2 µm. Coarse porosity is caused by the incomplete filling of interstices between previously deposited particles while

fine porosity is associated with the incomplete contact
between layers during the process of coating formation[6,7].

The coating microstructure also exhibits a fraction
of oxide stringers. The oxide content in the coating was
measured to be 9.2%. These oxides were formed as a result
of the reaction between the atmospheric air and the melted
particles during their flight from the plasma gun to the
substrate.

Figure 2 shows a scanning electron micrograph of the
transverse section of molybdenum coating on the steel
surface. Figures 3(a) and 3(b) show the X-ray images of
iron and molybdenum corresponding to Figure 2. As can be
observed, iron is not detected in the plasma sprayed
coating nor molybdenum in the substrate. In other words,
no inter-diffusion of the elements occurred between the
substrate and the coating during spraying. The diffusion
depth x is given by the following formula:

$$x = erf^{-1}(2\sqrt{Dt}) = \sqrt{Dt} \tag{1}$$

where D is the diffusion coefficient and t is the time.
It may be assumed that at the interface, the steel

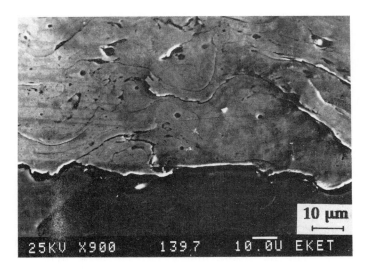

Figure 2 Scanning electron micrograph of the transverse
         section of the molybdenum plasma sprayed coating

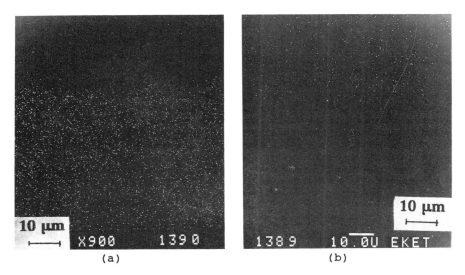

Figure 3 (a) X-ray image analysis of iron corresponding to Figure 2, (b) X-ray image analysis of molybdenum corresponding to Figure 2

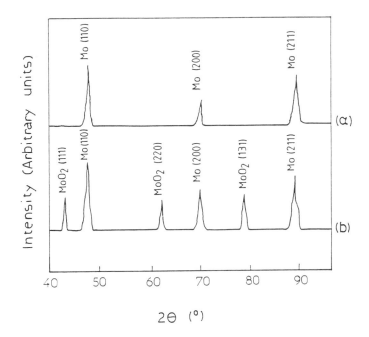

Figure 4 X-ray diffraction patterns of molybdenum before (a) and after (b) plasma spraying

substrate is heated to an intermediate temperature of
1200 - 1800^{0}C.   At these temperatures the diffusion
coefficient of molybdenum in iron varies from 4.8 x 10^{-12}
to 1.8 x 10^{-9} m^2/sec [8] and therefore the diffusion depth of
molybdenum in iron is expected to be 12-25 Å, which is
near the resolution of the SEM.   From the above
observations it may be concluded that the bonding of
molybdenum to the steel substrate is purely mechanical.

Figure 4 shows the X-ray diffraction patterns of
molybdenum before (a) and after (b) plasma spraying.   From
this Figure the molybdenum oxide formation during the
spraying process can also be observed.

After immersing each of the coated and uncoated steel
specimens in the HCl solution (pH=1.5) the change of the
corrosion potential was monitored as a function of
immersion time.   Electrode potentials were measured with a
high impedance data acquisition system.   The corrosion
potential of the uncoated steel was approximately -0.61 V

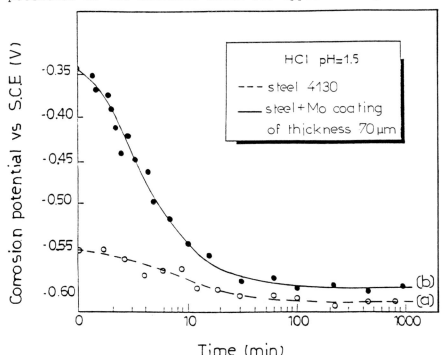

Figure 5 Corrosion potentials of (a) uncoated steel and
         (b) molybdenum plasma spray coated steel in HCl
         solution

after 10 min. and was essentially constant throughout the immersion period (curve (a) of Figure 5). The behaviour of the coated steel was similar with the potential falling from its initial value towards more active potentials asymptotically approaching the equilibrium corrosion potential curve (curve (b) of Figure 5). The corrosion potential of the coated steel after 1000 min. was -0.58 V or 30 mV more noble than that of the uncoated one. The corrosion potential of molybdenum coated steel is a mixed potential resulting from localized corrosion of the steel substrate beneath the porous plasma sprayed molybdenum coating.

The possible corrosion mechanism of the molybdenum coated steel can be considered as follows. The electrolyte initially penetrates through interconnected pores and attacks the substrate. The regions beneath those pores are exposed to the electrolyte which contains dissolved oxygen while neighbouring areas are leaner in oxygen. Therefore, due to the principle of differential aeration, the dissolution of the substrate starts at the regions adjoining the connected pores[9].

In acidic solutions the reactions governing the corrosion of steel are the following[10]:

$$Fe + Cl^- + H_2O \rightarrow (FeClOH^-)_{ads} + H^+ + e^- \qquad (2)$$

$$\overset{\cdot}{(FeClOH^-)}_{ads} \rightarrow FeClOH + e^- \qquad (3)$$

$$FeClOH + H^+ \rightarrow Fe^{2+} + Cl^- + H_2O \qquad (4)$$

The dissolved Fe^{2+} ions are unstable and transform to Fe^{3+} ions[9]:

$$2Fe^{2+} + 4(OH^-) + H_2O + 1/2O_2 \rightarrow 2Fe^{3+} + 6(OH^-) \rightarrow Fe_2O_3\,3H_2O \qquad (5)$$

Hematite Fe_2O_3 diffuses through the connected pores and accumulates on the coating surface forming light brown spots which can be seen in Figure 6.

The increased corrosion resistance of coated steel in comparison with the uncoated one may be attributed to the high corrosion resistance of molybdenum in the presence of chloride anions. In particular, the molybdate anions (MoO_4^{2-}) form, in the presence of Fe^{2+}, a passive layer of $FeMoO_4$ which prevents the easier incorporation of chloride anions in the coating and its easier dissolution[11].

<u>Figure 6</u> Plan view of the molybdenum coating surface after
one day of corrosion testing

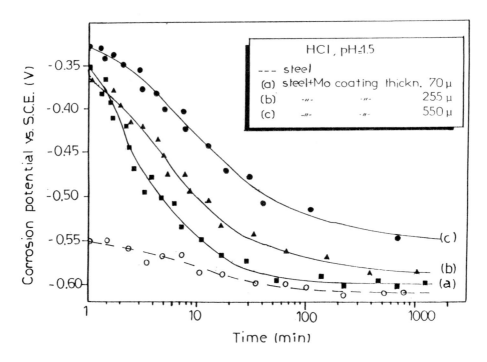

<u>Figure 7</u> Corrosion potential curves of steel coated with
molybdenum of various thicknesses

Figure 7 shows the corrosion potentials of coated steel with various molybdenum coating thicknesses. From this Figure it can be seen that the thicker the coating is the better the corrosion behaviour the coated system has. This phenomenon may be attributed to the fact that the number of interconnected pores decreases with increasing coating thickness. Consequently, the chloride anions find less paths to attack the steel substrate through the molybdenum coating.

Finally, Figure 8 shows the change in the weight of dissolved iron in the solution during the corrosion experiments of coated and uncoated steel. From this Figure it can be observed that the molybdenum coating decelerates the dissolution of the substrate thus improving the corrosion resistance.

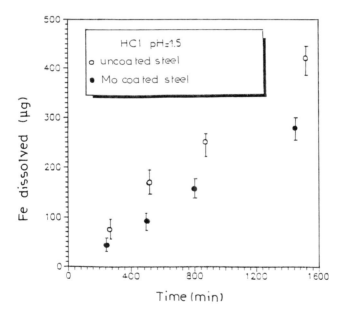

<u>Figure 8</u> Dissolved iron in the solution during the corrosion experiments of coated and uncoated steel

4 CONCLUSIONS

The corrosion behaviour of plasma sprayed molybdenum on steel was studied. The main conclusions of this study are given below.

1. During the plasma spraying process no elemental interdiffusion was observed between the coating and the substrate.

2. The corrosion of the coated specimen in HCl solution (pH = 1.5) was found to be lower than that of the uncoated specimen.

3. Increasing the thickness of the molybdenum coating improves the corrosion resistance of the coated system.

REFERENCES

1.   T.E. Strangman, Thin Solid Films, 1985, 127, 93.
2.   G. Johner and K. Schweitzer, J. Vac. Sci. Technol., 1985, A3, 2516.
3.   A. Levy and N. Jee, Wear, 1988, 121, 363.
4.   J. Ayers, R. Schaeffer, F. Bogar and E. McCafferty, Corrosion, 1981, 37, 55.
5.   R. Tucker, 'Deposition Technology for Films and Coatings', Noyes Publications, New Jersey, 1982.
6.   McPherson, Surf. & Coat. Technol., 1989, 39/40, 173-181.
7.   McPherson, Thin Solid Films, 1981, 83, 48.
8.   J. Houben, G. van Liempd, Proc. 10th Int. Thermal Spraying Conf., Essen, DVS-Verlag, 1983, 66-71.
9.   J. Bockris and A. Reddy, 'Modern Electrochemistry 2', Plenum, New York, 1977.
10.   K. Rajagoplan and G. Venkatachari, Corrosion, 1980, 36,(6), 155.
11.   M. Stranick, Corrosion, 1984, 40(6), 296-302.

1.5.5
Subplate Effects upon the Porosity of Electrodeposited Gold

G. W. Marshall,[1] S. V. Allen,[2] M. Tonks,[1] and D. Clayton[1]

[1] SCHOOL OF ENGINEERING, SHEFFIELD HALLAM UNIVERSITY, SHEFFIELD SI IWB, UK

[2] GOLDRITE LTD, SHEFFIELD, UK

1   INTRODUCTION

Corrosion products formed on the surface of electrodeposited gold originate from either impurities within the gold coatings or more often as a result of corrosion of the substrate metal through pores, or discontinuities, within the gold surface. It is therefore of the utmost importance to eliminate, or reduce to a minimum, the porosity of gold deposits if they are to show maximum corrosion resistance during their service life.

Historically the measurement of porosity and studies of pore formation in electrodeposits has been associated with nickel, chromium and gold plating processes. The subject was reviewed generally by Clarke[1] and more specifically with respect to electrodeposited gold coatings by Garte[2].

The porosity of gold deposits is affected by many factors including: surface roughness, preplate treatment of the substrate, plating bath composition and operating characteristics, plate thickness and post plate treatment of the coatings. Many techniques have been used to measure porosity including a gaseous test developed by Clarke and Leeds[3]. This test is straight forward, easy to carry out, and can be used to study the porosity of gold deposits on a number of substrates. Many of the results of past work are confusing and often appear contradictory, due no doubt to interaction of the many factors affecting pore formation. The gold content of commercial plating baths has been reduced progressively as gold prices have increased. Thus many results obtained during earlier studies on the porosity of gold deposits do not apply to gold deposits produced using present practices.

The allergenic nature of nickel is causing concern over its use as a subplate during the gold plating of items which might be subject to contact with human skin. Therefore there is considerable interest in alternative coatings, which might be used as substitutes for nickel subplates during gold plating. Possible alternatives are copper and bronze deposits.

The former has two disadvantages, namely, their colour and their
tendency to diffuse into the gold overplate[4,5].

The object of this work was to compare the porosity of thin
gold deposits overplated onto different subplates, namely nickel
copper and bronze. It formed a natural extension to earlier
work aimed at determining the affects and interaction of the
operating parameters on the plating efficiency of a bright
gold-nickel alloy plating bath[6].

2   EXPERIMENTAL WORK AND RESULTS

Preparation and Preplating of Test Panels

Standard brass Hull cell test panels (W Canning Materials
Ltd.) were cut into two pieces, each measuring approximately 51
x 76 mm. Two holes, 3mm in diameter, were drilled in two cor-
ners adjacent to one of the short sides of each panel to facilitate their suspension in the plating bath during efficiency
measurements.

Preplating of Test Panels. The test panels were preplated
with nickel, copper or bronze subplates prior to gold plating.
In each case the preplating sequence was similar and involved
degreasing, washing, electrolytic cleaning and washing prior to
the formation of the subplate coatings using the following
proprietary baths.

A 'Cupracid^R 210' (Shering A G) bath operated at 30°C and
a cathode current density of 2 A dm^{-2} with suitable agitation
was used to form bright copper subplates 6 μm thick.

A 'Udylite^{TM}66' (OMI International) bath operated at 60°C
and a current density of 4 A dm^{-2} with air agitation was used to
form bright nickel subplates 6 μm thick.

An 'Imitor 2000' (Schloetter Co. Ltd.) bath operated at
45°C and a cathode current density of 0.4 A dm^{-2} with mechanical
agitation was used to form bright bronze subplates 5 μm thick.
The deposits consisted of mainly copper with smaller amounts of
zinc and tin. They also contained some 1% lead arising from the
brightening system used.

Examination of the subplates using the SEM showed that the
nickel subplate surfaces were smoother than those of the copper and
bronze subplates. The latter surfaces had rougher textured
appearances, similar looking to flat sand or orange peel surfaces,
when observed at x1000 - 2000 magnification.

All preplated test panels were washed well following plating. Nickel and copper plated panels to be used in the measurement of cathode current efficiency during gold plating were
dried and stored in a desiccator until required.

Measurement of Gold Plating Efficiencies

Experimental Procedure. Nickel and copper preplated test panels

were held in a purpose made jig during electrolyses carried out
to determine the cathode current efficiency for gold plating.
The jig's surface was stopped off using a heat curing epoxy
resin and only those small areas used to make electrical contact
were left free. During gold plating experiments the test panels
were immersed in 4l of plating solution and positioned parallel
to, and mid distance between two graphite anodes. Agitation of
the bath was maintained by passing air at a controlled rate
through a perforated coiled tube positioned below the cathode.
The current used for individual electrolyses was preset using a
'dummy' test panel. The plated coatings were activated prior to
use as described below.

The plating bath. The plating bath used for these experiments was a proprietary one 'Endura Gleam 205' (Lea Ronal)
having gold and nickel contents maintained in the ranges 2.7 -
2.8 and 4.75 - 5.15 g/l respectively and operated at a temperature of 45°C and within a pH range of 4.1 - 4.2.

The Plating Efficiencies. Plating efficiencies were calculated for gold deposition using the data obtained from the above
experiments. Compositional data previously obtained for gold-
nickel deposits obtained for the bath on copper substrates was
used in these calculations[6]. In the absence of any data
relating to the composition of gold-nickel deposts obtained from
the bath on nickel substrates it was assumed that they had the
same composition as those obtained on copper substrates under
similar plating conditions.

The current efficiencies for gold alloy deposition onto
copper and nickel substrates using current densities of 0.5 and
1.0 A dm^{-2} are given in table 1. The efficiency observed for
gold deposition on copper subplates was greater than that found
for deposition onto nickel subplates.

The current efficiency for gold alloy deposition onto nickel
subplates as a function of plating time at a current density of 1.0 A
dm^{-2} is shown in figure 1. The current efficiency tends to increase
with increasing plating times up to about three and a half minutes
after which it reaches a plateau.

Table 1 The current efficiency for gold alloy deposition onto nickel and
copper sulphates

Subplate	Current Density (A dm^{-2})	Plating Time (S)	Current Efficiency (%)
Cu	0.5	180	28.4
Cu	1.0	90	24.4
Ni	0.5	180	19.4
Ni	1.0	180	20.4

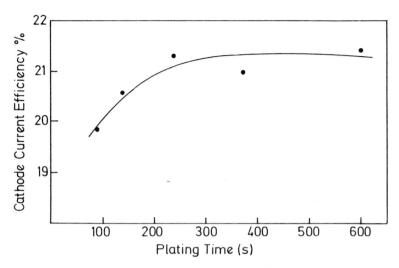

<u>Figure 1</u> Variation of current efficiency with plating time for gold
plating onto nickel substrates

Porosity Measurements

<u>Gold Plating of Test Panels.</u> Following preplating and
washing, the test panels were given the following pre-treat-
ments.

Nickel plated test panels were cathodically activated using
Lea Ronal Activator No. 2 solution.

Copper plated panels were etched in a solution of Metex 629
(MacDermit, Telford) dry acid salts.

Bronze plated test panels were given the same pretreatment
as the copper plated panels.

The pre-treated panels were washed and immediately gold
plated using a proprietary bath, 'Endura Gleam 205' using a
current density of 0.8 A dm^{-2}. The bath was operated at 45°C
and a pH of 4.1 with vigorous air agitation. The gold and
nickel contents of the bath were maintained within the ranges
2.6 to 2.75 and 4.64 to 4.8 g l^{-1} respectively.

The gold plated panels were first washed in cold then hot
water before being dried using hot air.

<u>The Test Panels</u>. Three sets of three test panels preplated
with nickel, copper and bronze respectively were overplated with
0.25 μm of gold for use in porosity tests.

A further series of four panels with nickel subplates were overplated with thicknesses of 0.25, 0.5, 0.75 and 1.0 μm of gold respectively. This series was duplicated using four panels preplated with copper rather than nickel subplates.

Measurement of Deposit Thickness. The thicknesses of nickel, copper and bronze subplates were checked using a metallurgical microscope fitted with an eyepiece containing a calibrated scale. Metallurgical cross sections of mounted samples taken from the centres of plated panels were prepared for these measurements.

The thicknesses of gold on duplicate plated specimens was carried out using a Twin City coulometric thicknesses measuring instrument which was calibrated to give a thickness reading directly in microns. These thickness measurements were made at two points near to the centres of the test samples.

All thickness measurements were within $\pm$ 8% of the values quoted.

Gaseous Porosity Tests. Plated samples were subjected to the gaseous porosity test at room temperature developed by Clarke and Leeds[3]. A desiccator having a volume of 2.35 l served as the test chamber. The plated test pieces were held in slits cut in a perspex holder which supported them in the main chamber above 58.75 ml of test solution held in a dish within the well of the desiccator (figure 2). The test solution was made up by mixing four volumes of 20% sodium thiosulphate solution with one volume of a 50/50 concentrated sulphuric acid/water mixture. This produces an atmosphere of approximately 10% sulphur dioxide and a relative humidity of 86 percent when the solution volume is one fortieth of that of the chamber.

The gold plated test panels were placed in the test chamber and left for twenty four hours during which time brown spots developed indicating the positions of pore sites. The samples were removed from the chamber and examined for porosity as soon as was practically possible, within 24 hours. It was subsequently noted that the corrosion products forming the brown spots began to spread significantly two or three days after the removal of samples from the chamber.

Pore Counting. All the 'exposed' test panels were examined on the smooth side using a mask (figure 3) to define two rectangular areas, 20 x 25 mm. Pore counts were made using a microscope, fitted with a Seascan image analyser which has inbuilt computerised counting facilities based upon image density. Pore counts were made over the two areas 'a' and 'b' defined by the mask on each panel tested.

It was noted that plates appeared to be more porous on the rough sides as compared to the polished sides of the panel although no quantitative measurements were made relating to this observation.

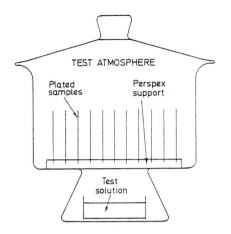

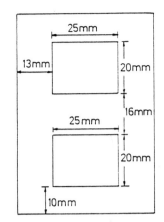

Figure 2 **Gaseous Porosity Test Apparatus**   Figure 3 Mask used for Porosity
                                                        Counts on Two Areas Each 2x2.5 cm

Results of Pore Counts. The porosity counts made on 'exposed' test
panels with 0.25 μm of gold overplated onto nickel, copper and bronze
subplates respectively are given in table 2. The results give an
average pore density for gold on nickel, copper and bronze subplates
of 28.9, 5.2 and 9.5 pores cm^{-2} respectively.

The porosity of gold overplates on nickel and copper subplates as a function of gold thickness are shown in figure 4.

Table 2 The porosity of 0.25 µm thick fold plates on nickel, copper
and bronze subplates

Sample	Pore Count (5 cm^{-2})	Sample	Pore Count (5 cm^{-2})	Sample	Pore Count (5 cm^{-2})
1 Ni	a) 203 b) 133	1 Cu	a) 13 b) 38	1 Bronze	a) 47 b) 57
2 Ni	a) 149 b) 137	2 Cu	a) 20 b) 15	2 Bronze	a) 37 b) 49
3 Ni	a) 128 b) 94	3 Cu	a) 26 b) 20	3 Bronze	a) 68 b) 27
4 Ni	a) 126 b) 188	4 Cu	a) 40 b) 35		
Total	1158		207		285
	28.9		5.2		9.5
Porosity (Average Counts cm-2)					

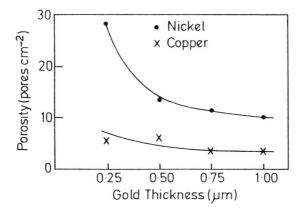

The porosity of gold deposits as a function of thickness on
          nickel and copper subplates

It can be seen that the porosity of gold overplates was
greater on nickel than copper subplates for all thicknesses of gold
tested. Furthermore the porosity of gold coatings decreased as
their thickness increased when overplated onto nickel or copper.

In all cases test panels plated with nickel, copper or
bronze but having no gold overplate became brown all over after 2
or 3 hours when placed in the porosity test chamber.

3   DISCUSSION

Current Efficiency Measurements

Two aspects of current efficiency measurements are of
interest: (a) the lower efficiency observed when gold was
deposited onto nickel rather than copper, and (b) the increase
in current efficiency with increasing plating times, up to about
three and a half minutes, when gold is overplated onto nickel
subplates. An explanation for both phenomena may be linked to
Piontelli's classification of metals in accordance with their
electrochemical behaviour[7]. Thus metals are divided into three
groups, 'normal' metals including copper, gold, tin, zinc and
lead, 'intermediate' metals, and 'inert' metals including nickel. 'Normal' metals are characterised by the relative ease with
which they are deposited from solution and in contrast the
relative difficulty of hydrogen discharge at their surfaces.
This situation is reversed in the case of 'inert' metals.

Hence one might expect less competition to gold discharge
from hydrogen discharge during the initial stages of gold plating onto a copper surface (ie. a 'normal' metal) than when
plating onto a nickel surface (ie. an 'inert' metal). Such a
situation would possibly explain the observed higher efficiency
found for gold plating onto copper rather than nickel surfaces
following relatively short plating times.

Similarly during the initial stages of gold plating onto
nickel surfaces one might expect greater competition from the
co-discharge of hydrogen than in the latter stages once the
original nickel surface (ie. an 'inert' metal surface) has been
replaced by a gold surface (ie. a 'normal' metal surface).
Furthermore the cathode current efficiency achieved for electrolyses
lasting longer than some four minutes probably represents the 'steady
state' efficiency for continued growth of the gold-nickel alloy
deposit.

Porosity of Gold Deposits

Porosity and Deposit Thickness. It is well established that
the porosity of electrodeposits decreases with their increasing
thickness provided their roughness does not vary greatly. Thus
the decrease in the porosity of gold deposits with increasing
thickness on both nickel and copper subplates is in line with
earlier work.

Porosity and Subplates. The situation is not so clear cut
when considering the affects of nickel and copper subplates on
the porosity of overplated gold. Hence Ashurst and Neale found
that nickel underplates were more effective than those of copper
in reducing the porosity of overplated gold[8]. However Garte's
work on the affect of nickel and copper subplates, from Watts
and acid baths respectively, on the porosity of 0.75 μm thick
gold deposits probably correspond nearest to the present work as
do the results[9]. Both this earlier work and the present study
showed copper underplates from acid baths to be more effective
in reducing the porosity of thin gold deposits than nickel
underplates from Watts type baths. It is worth noting that the
differences in porosity found in the present work cannot be
explained in terms of the substrate surface roughness. On the
basis of surface roughness alone it might have been expected
that the porosity of gold deposits on the smoother nickel subplates would have been less than those on the rougher copper and
bronze subplates. In practice the opposite was found to be the
case.

Mechanism and Pore Formation. It is generally held that
'pore precursors'[1,2,9] in the form of surface defects[10] play a
major role in pore formation. These may take the form of surface inclusions, voids, crystallographic defects or mechanical
stress points. The role of 'pore precursors' in pore formation
is to inhibit the normal initial growth of the electrodeposited
metal, via nucleation rapidly followed by lateral
growth[11,12,13].

The role of hydrogen co-deposition during gold plating on
the porosity of the coatings has received relatively little
attention. Leeds and Clarke reported some efficiency measurements but these related to baths containing much greater gold
concentrations than those used in the present work.

It is reasonable to suppose that the relatively high incidence of hydrogen discharge during the initial stages of gold
plating onto nickel surfaces increases the surface density of

'pore precursors' on the surface. This increase in the density
of 'pore precursors' might result from either an enhancement of
the effect of existing 'pore precursors' or the development of
new precursors resulting from high hydrogen co-deposition. Thus
hydrogen bubbles might form at the sites of relatively small 'pore
precursors' which, in the absence of the discharged hydrogen, might
have been rapidly bridged by the growing metal deposit. Ogburn and
Ernst have shown that hydrogen bubbles can play such a role in the
formation of pores during nickel plating[14].

Practical Implications. Whatever the reason for the observed lower porosity of gold deposits on copper or bronze as
compared to those on nickel subplates the result is of practical
significance. The best answer to achieving low porosity gold
deposits may depend upon the function they are to perform in
service. Replacing the usual nickel subplate with a bronze one
might be the best course where articles such as spectacles and
watches which will come into contact with human skin are to be
gold plated. This would overcome any problems associated with
the allergenic properties of nickel. However it may be better
to use a combination of a relatively thick nickel underplate in
conjunction with a flash bronze subplate for the electronic
applications of gold plating. This would take advantage of both
the levelling properties of nickel deposits and the ability of a
bronze subplate to produce low porosity gold deposits.

ACKNOWLEDGEMENTS

The authors wish to thank Goldrite Ltd, Sheffield and Schloetter
Company Ltd, Pershore, for assistance in carrying out this work.

REFERENCES

1. R. Sarde, 'Properties of Electrodeposits', The Electrochemical
   Society Incorporated, Princeton, 1975, Chapter 8, p.122-140.
2. F.H. Reid and W. Goldi, 'Gold Plating Technology', Electrochemical Publications Ltd., Teddington, 1974, Chapter 23,
   p.295-315.
3. M. Clarke and J.M. Leeds, Trans. Inst. Metal Finishing,
   1968, 46, 81.
4. M.S. Frant, Plating, 1961, 48, 1305.
5. M.R. Pinnel and J.E. Bennet, Proc. Holm. Seminar on
   Electrical Contact Phenomena, Illinois Inst. of Technology,
   Chicago, 1971.
6. M. Tonks, BEng Project Thesis, Sheffield City Polytechnic, 1990.
7. R. Piontelli, CITCE2, Tamburino and Milan, Butterworths,
   London, 1951.
8. K.G. Ashurst and R.W. Neale, Trans. Inst. Metal Finishing,
   1967, 45, 75.
9. S.M. Garte, Plating, 1968, 55, 946.
10. G.L. Cooksey and H.S. Campbell, Trans. Inst. Metal Finishing,
    1970, 48, 93.
11. G.I. Finch and C.H. Sun, Trans. Faraday Soc., 1936, 32, 852.
12. G.I. Finch and A.L. Williams, J. Electrodepositors Tech.
    Soc., 1937, 12, 105.

13. G.I. Finch, H. Williams, and L Young, Disc. Faraday Soc., 1947, 1 44.
14. F. Ogburn and D.W. Ernst, Plating, 1959, 46 831, 957.

1.5.6
The Relationship between Steel Surface Chemistry and Galvanising Reactivity

R. W. Richards,[1] H. Clarke,[1] and F. E. Goodwin[2]

[1] DIVISION OF MATERIALS AND MINERALS, UNIVERSITY OF WALES, CARDIFF CF2 1XH, UK

[2] MATERIALS SCIENCES, INTERNATIONAL LEAD ZINC RESEARCH ORGANIZATION, PO BOX 12036, NORTH CAROLINA, USA

1   INTRODUCTION

Batch or job hot dip galvanising is an efficient and economical means of imparting corrosion resistance to a wide range of prefabricated steel articles. Typical examples include much of the steelwork used in civil engineering works (e.g. girders, beams, box sections etc.), street furniture (e.g. lamposts, crash barriers, fencing, etc.) and general industrial and agricultural plant parts. Frequently, such articles are combined with sections produced from hot dip galvanised steel sheet (produced on continuous steel sheet coating lines) this giving the final product a much extended corrosion-free life span.

The process of job galvanising has changed little over the years. The prefabricated article (frequently of complex construction from several grades of native hot-band steel) is initially degreased and then pickled in hydrochloric or hot sulphuric acid. These operations remove in sequence, ubiquitous organic surface contaminants such as oils and grease, and the often thick surface oxides which would otherwise prevent wetting of the steel surface by liquid Zn. The pickling stage is often followed by an aqueous rinse before fluxing in an aqueous solution of zinc and ammonium chloride salts. On withdrawal, the flux solution evaporates leaving a thin crystalline deposit of fluxing salts, this enabling uniform wetting of the steel by the molten Zn. The steelwork is then immersed in a bath of commercial purity liquid Zn at temperatures in the range 445-460°C for times of the order 4-7 minutes.

On immersion in the melt the steel reacts with the Zn forming a layer of brittle Fe-Zn intermetallic compounds (viz., the 'alloy' layer). Three phases of decreasing Fe content are present within this layer. These are the Γ, δ and ζ phases respectively. The Γ

phase, adjacent to and contiguous with the substrate, is often too thin to be observed. The δ phase has two distinct morphologies: a compact layer adjacent to the Γ phase, extending to a pallisade or discontinuous structure adjacent to ζ. The ζ phase often comprises finely dispersed particles within a matrix of pure Zn, a final outer pure Zn coating having been deposited on the ζ layer when the article is withdrawn from the melt.

The typical coating structure is shown in Figure 1. The thickness of the outer pure Zn layer depends on bath viscosity, speed of withdrawal, and the reactivity of the steel, since it is possible that this layer is consumed in further alloy layer growth after withdrawal.

Reaction diffusion between Zn and pure Fe leading to Fe-Zn intermetallic formation is nominally a volume-diffusion controlled reaction displaying classical, parabolic growth kinetics.[1,2] However, the presence of other elements in steel can modify alloy layer growth kinetics so that much faster rates of alloy formation occur, these better described by linear rate laws. Although the mechanistic reasons for this remain obscure[3] the overall effect is to produce excessively thick alloy layers which predispose the coating to mechanical damage and delamination; it is also wasteful of Zn.

The two steel constituents most likely to expedite Fe-Zn reaction diffusion and lead to excessively thick coatings are Si and P.

The effect of substrate Si was first quantified in a detailed study by Sandelin[4] who showed that a peak in Si-induced high reactivity characterised by excessively thick alloy layers occurred with steel Si concentrations of about 0.10 wt.%. A subsequent decrease in reactivity with Si concentrations of about 0.15-0.3 wt.% was found to be followed by an increase in reactivity as the Si level approached 0.4 wt.%. His results were largely confirmed by Horstmann and other workers[5-8] who generally summarise the Sandelin Si effect as shown in Figure 2: this shows a peak in alloy thicknesses within a substrate Si concentration range of about 0.06 to 0.15 wt.%.

Subsequently, several further studies have shown that substrate P acts synergistically with Si, both elements expediting the rate of reaction and leading to excessively thick coatings.[9-12] The results of these studies have been expressed in terms of the Si-P galvanisability graphs shown in Figures 3 and 4[13,14] where substrate Si and P concentrations likely to

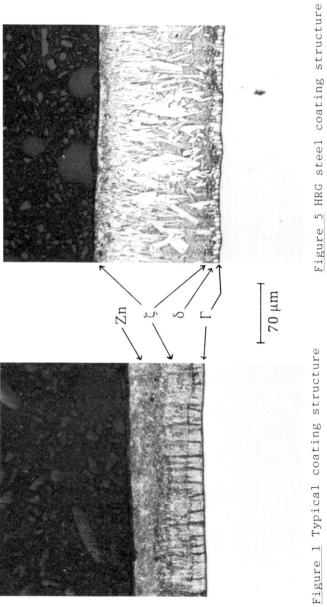

Zn

ζ

δ

Γ

70 μm

Figure 5 HRG steel coating structure

Figure 1 Typical coating structure

Figure 2 Hot dip galvanising: effect of Si - the "Sandelin"
         reactivity trend

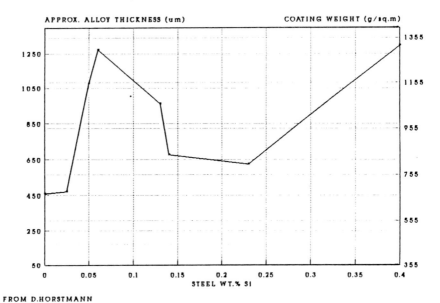

FROM D.HORSTMANN

Figure 3 French Si:P galvanising standard, Reference 13

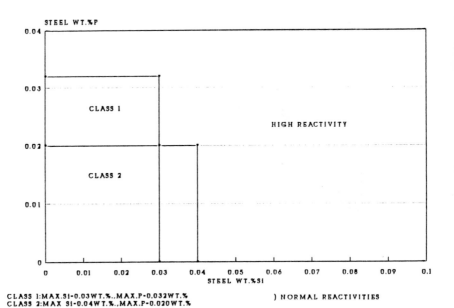

Figure 4 German Si:P galvanising standard, Reference 14

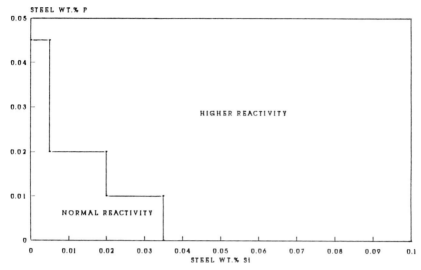

MAXIMUM SI-P RATIOS IN WT.% ARE:
(0.005SI.(0.045P.; (0.020SI.(0.020P.;
AND (0.035SI.(0.010P

Figure 6 Alloy layer thickness in steels 1-5 (see Table 2) after dipping for 6 mins. at 455^{0}C

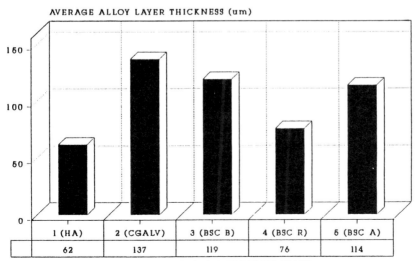

induce high reactivity are seen to occur at Si levels in excess of about 0.03-0.04 wt.%, these combined with P concentrations in excess of about 0.01-0.02 wt.%.

The structure of a typical Si-P induced thicker coating is shown in Figure 5.   It shows a depleted δ phase layer underlying an excessively thick ζ phase layer of finely dispersed particles within a pure Zn matrix.

One frequently argued cause for Si-induced high reactivity is enrichment of the steel surface by Si, presumably occurring during hot-rolling of the steel. Although unequivocal experimental evidence for the influence of substrate surface silicon on galvanising reactivity is lacking, several groups of workers have demonstrated the apparent importance of surface chemistry on galvanising reactivity.[15-17]

This work reports the salient aspects of an extensive study on the relationship between galvanising reactivity and substrate chemistry.   The work has involved the identification of reactive substrates and subsequent correlation of reactivity data with bulk and surface chemistries.

2   EXPERIMENTATION AND RESULTS

For this study over 50 commercial substrates were obtained from galvanising industry sources.   The general range of C, Mn and tramp element concentrations for these steels is shown in Table 1.   Table 2 shows specific data for five important representative substrate samples.   All bulk compositional data was obtained by standard spark spectrometry analysis.   The specific Si and P concentrations of all steels will be reported later.

Table 1 General composition range of all steels

Constituent	General Range (wt.%)
C	0.040-0.392
Mn	0.142-0.426
Al	0.001-0.076
Si	0.001-0.391
P	0.003-0.080
S	0.002-0.041
Cr	0-0.049
Cu	0.002-0.041
Ni	0.008-0.056
Ti	0-0.009

<u>Table 2</u> Detailed compositions of five representative steels (wt.%)

Substrate	C	Mn	Al	Si	P	S	Cr	Cu	Ni
1.HA	.164	.130	.033	.004	.013	.015	.021	.017	.021
2.CGALV	.076	.360	.045	.017	.076	.015	.016	.015	.011
3.BSCB	.103	.390	.021	.105	.005	.004	.010	.033	.010
4.BSCR	.296	.371	.023	.157	.004	.013	.030	.012	.037
5.BSCA	.092	.360	.031	.360	.111	.007	-	.120	.013

From the parent steels, samples of dimensions 4cm x 2cm and of gauge < 2mm were cut and hot dip galvanised for 6 mins. at 455°C in a bath of commercial purity spelter. Pre-dip preparation consisted of degreasing in an aqueous degreasing agent, pickling in approximately 20% v/v hydrochloric acid for 5 mins., and fluxing in a saturated aqueous solution of zinc and ammonium chlorides. Dipping was performed using an automated immersion-withdrawal rig. Specimens were water quenched within 3 secs. of withdrawal from the melt to prevent post-dip reaction diffusion of the free Zn outer layer on specimen withdrawal. Specimens were then sectioned and transverse-sections hot-mounted in bakelite and prepared in accordance with standard procedures.[18,19] Alloy layer thicknesses were measured metallographically taking the mean of nine measurements per sample side.[19] The mass of data obtained is presented in Figures 6 and 7.

Figure 6 shows the variation in alloy thicknesses for the five important samples listed in Table 2. Figure 7 shows the general distribution of alloy thicknesses expressed in terms of the Si-P ratios for each steel. For the construction of this plot higher reactivity steels (Series 2 in the plot) have been arbitrarily defined as those steels which give mean alloy thicknesses of greater than 100μm after 6 mins. at 455°C. Normal or lower reactivity steels (Series 1 in the plot) have been defined as those steels which give a mean alloy thickness of less than 100μm under the same conditions. The mean thickness of the former series was found to be 124 ± 15μm whilst the analogous datum for the latter series was found to be 84 ± 10μm.

Figure 7 Correlation of alloy thicknesses with steel Si
and P content for all substrate samples

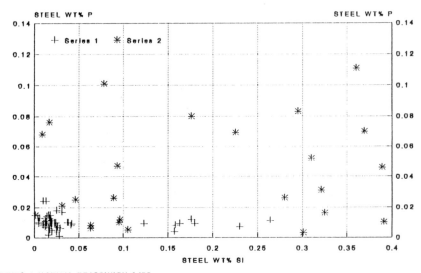

SERIES 1: NORMAL REACTIVITY < 100um
SERIES 2: HIGHER REACTIVITY > 100um

In order to evaluate the claim that higher reactivity steels have surfaces enriched with Si, X-ray photoelectron spectra (XPS) of the surfaces of a number of the steels were obtained using a VG Escalab Mk.II instrument. All spectra were obtained from substrates in a galvanisable condition, i.e. having been both degreased and pickled. Spectra were obtained at approximately 20nm depth intervals over analysis areas of about 2-10mm^2, each analysis pertaining to a penetration of about 0.2nm. Depth profiling was achieved using standard argon-ion beam etching.

Example spectra for the five representative steels are shown in Figures 8(a)-8(e), each montage showing the spectra obtained after successive 20nm etch steps down to approximately 180nm.

Further depth composition profiles for Si over much larger sample areas of 100cm^2 were obtained by a combination of surface chemical etching and atomic absorption spectroscopy analysis of etchant solutions using a technique analogous to that reported by Hudson et al.[20,21] Three example Si depth profiles produced by this method are shown in Figures 9(a)-9(c).

3   DISCUSSION

Reference to Table 2 shows detailed bulk compositional analyses for five representative experimental substrates (numbered 1-5). These five steels conveniently span the Si and Si and P composition ranges shown in Figures 2, 3 and 4 respectively.

On the basis of the Sandelin plot alone (i.e. Figure 2) only steels 3 and 5 would be expected to have enhanced reactivity, whilst steels 1, 2 and 4 would be expected to be of normal reactivity. If, however, reactivity is interpreted in terms of a combined Si and P content using Figures 3 and 4, it can be seen that only steel 1 should be of normal reactivity with steels 2-5 showing the greater reactivity expected of higher Si and P containing substrates.

However, as Figure 6 clearly indicates, the results of this work show that steels 1 and 4 are of lower reactivity than steels 2, 3 and 5, suggesting neither Figure 2 nor the more recent works represented by Figures 3 and 4 adequately describe enhanced steel reactivity due to the presence of Si and P.

A more detailed picture of Si and P induced higher reactivity can be gleaned from Figure 7. Here it is evident that the lower reactivity steels are generally confined to low combined Si and P concentrations over the whole range of Si compositions studied. Unlike the

<u>Figure 8</u> XPS depth profiles: (a) HA, (b) CGALV, (c) BSC B,
          (d) BSC R, (e) BSC A

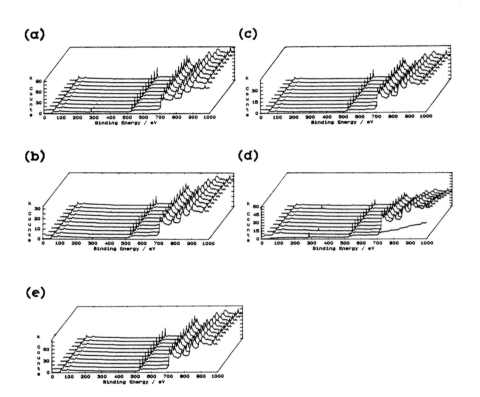

Sandelin plot in Figure 2 which suggests that only Si concentrations in excess of 0.06 wt.% are likely to effect higher reactivity regardless of concomitant P effect, Figure 7 clearly shows that low Si concentrations (i.e. < approx. 0.04 wt.%) remain subject to enhanced reactivity provided the P content is in excess of a certain limiting value. This limiting value of P increases as the Si content decreases, i.e. decreasing Si concentration demands an increasing P concentration to maintain the higher rate of reactivity. This is in agreement with the data of Figures 3 and 4.

However, Figure 7 also shows that it is possible to have relatively high Si steels with low P concentrations that react normally, in contradiction of the data shown in Figures 3 and 4.

This latter finding tends to concur with Sandelin's results which indicate that a region of normal reactivity lies between Si concentrations of about 0.15-0.30 wt.%. On the basis of this study, reactivity trends in the region of the Sandelin reactivity peak (i.e. 0.06-0.15 wt.% Si - see Figure 2) remain uncertain since the number of commercial substrates with Si and P concentrations in this range available for study has been limited. However, it seems reasonable to assume that Si containing steels in this region with low or negligible P concentrations may react normally. This would lie in agreement with Böttcher's assertion[22] that not all previous research has unequivocally confirmed the presence of the Sandelin reactivity peak. This situation will be clarified when further commercial substrates with a range of Si-P concentrations in the region of the Sandelin peak become available for study.

On the basis of the results obtained so far, it can be tentatively asserted that Si-P induced higher reactivity is best represented by a curve of the type shown in Figure 10. This curve attempts to establish a borderline between normal and higher reactivity steels on the basis of the results shown in Figure 7, although in practice a zone of intermediate reactivity is likely to exist. Despite the limitations of such a representation, such a plot may be of use to industrial galvanisers in attempting to predict higher reactivity steel compositions.

XPS analyses of the surfaces of steels 1-5 after pickling are shown in Figures 8(a)-8(e). Before analysis, remaining oxide scale was removed by argon-ion beam etching so that each successive spectrum pertains to 20nm depth intervals from the metal surface. Identification of the scale-metal interface

was deduced from the change in the form of the XPS plot
on etching;   this is clearly seen in Figure 8(d) for
steel 4.   The XPS instrument used is capable of
detecting   and   quantifying   elements   down   to
approximately 0.1 At.%.   If enrichments of Si were
present in the substrate surfaces a quantifiable peak
would be clearly seen at a binding energy value on the
abscissae of the plots at about 103 eV.   In none of the
spectra, at any depths down to 180nm, is such a peak
observed.   The implication of these results is self-
evident: in none of the steels is there evidence for
Si   enrichment   of   the   surface.   The   absence   of
quantifiable Si peaks in even the higher Si containing
steels   could   be   interpreted   as   evidence   for   Si
depletion of the metallic surface.   This could well
result from preferential oxidation of substrate surface
Si during hot-rolling and its subsequent loss as mill-
scale.

Further evidence for inhomogeneous distribution of
Si in the substrate surfaces is provided by the AAS-
chemical   etching   analyses   described   previously.
Laboratory analyses suggest that this technique is
capable of detecting variations of ± 0.03 wt.% Si at a
95% confidence level.   Three example depth profiles
for Si determined by this method are shown for steels
3, 4 and 5 (Figures 9(a)-(c)).   Inspection of these
data again suggests low Si values at the surface as
compared with average bulk values determined by spark
spectrometry.   The variation in Si value at various
depths also suggests that the distribution of Si at the
near surface is subject to some variation although
further work using different surface analysis methods
is currently being undertaken.

On the basis of the surface analysis results
obtained, it seems most improbable that substrate Si
surface enrichments are the source of Si-P induced high
reactivity phenomena as reported previously.

4   CONCLUSIONS

Previously   published   data   on   Si-P   induced   high
reactivity galvanising inadequately quantifies the
nature of the phenomena over the whole composition
range of Si and P containing steels currently
encountered by the job galvaniser.   This work suggests
that excessively thick coatings are produced on steels
providing the P content exceeds a certain limiting
value dependent on residual Si content.

Enrichment of the substrate surface by Si is not a
plausible explanation of Si induced high reactivity.

<u>Figure 9(a)</u> AAS depth profile - steel 3 (BSC B), Si depth
                profile, bulk CONC.Si = 0.106 wt.%

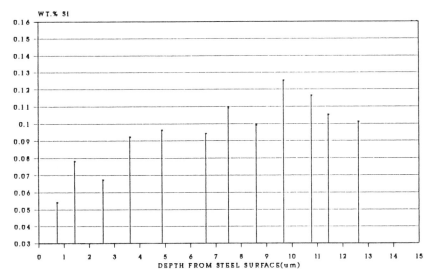

<u>Figure 9(b)</u> AAS depth profile - steel 4 (BSC R), Si depth
                profile, bulk CONC.Si = 0.158 wt.%

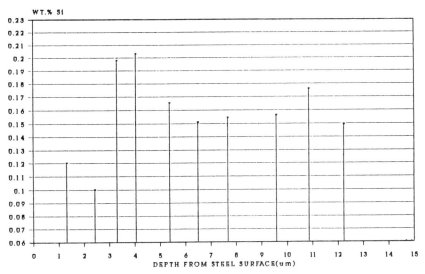

Figure 9(c) AAS depth profile - steel 5 (BSC A), Si depth
           profile, bulk CONC.Si = 0.360 wt.%

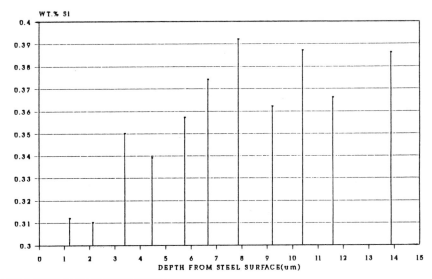

* BULK VALUE BY SPARK SPECTROMETRY

Figure 10 Suggested boundary between normal and higher
          reactivity steels

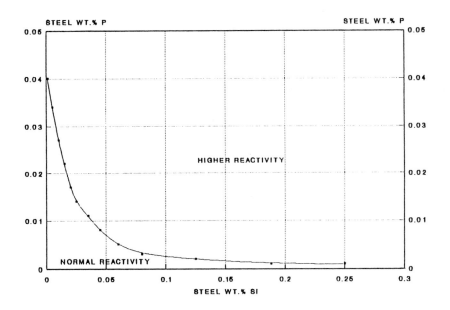

ACKNOWLEDGEMENTS

Richards and Clarke gratefully acknowledge the funding of this project by the International Lead Zinc Research Organization, Research Triangle Park, North Carolina, U.S.A. Thanks are also due to Hereford Galvanisers Ltd., Cardiff Galvanizers Ltd., and Pasminco Europe for the provision of various experimental materials.

REFERENCES

1. D. Horstmann, Reactions between Iron and Molten Zinc, Zinc Development Association, London, 1978.

2. J. Makowiak and N. R. Short, *Inter. Mets. Revue.*, 1979, Review 237, p.1.

3. P.J. Gellings, in Procs. of Seminar: The Galvanizing of Silicon Steels, Liege, Belgium, May 1985, International Lead Zinc Research Organization, Research Triangle Park, North Carolina, U.S.A.

4. R.W. Sandelin, *Wire and Wire Prods.*, Dec. 1940, 15, p.721.

5. D.C. Pearce, Ibid., reference 3.

6. D. Horstmann and F.K. Peters, in Procs. of 9th International Galvanizing Conf., Dusseldorf, 1970, Zinc Development Association, London, p.75.

7. J.J. Sebisty and R.H. Palmer, in Proc. of 8th International Galvanizing Conf., London, 1967, Zinc Development Association, London, p.30.

8. W. Warnecke, L. Meyer and A. Selige, in Procs. of 13th International Galvanizing Conf., London, 1982, Zinc Development Association, London, p.31/1.

9. J. Pelerin, J. Hoffmann and V. Leroy, *Metall. und Technik*, 1981, 35, 9, p.870.

10. D. Horstmann, *Arch. Eisen.*, 1975, 46, 2, p.137.

11. G. Hansel, in Procs. of 12th International Galvanizing Conf., Paris, 1979, Zinc Development Association, London, p.163.

12. G. Hansel, *Metall. und Technik*, 1980, 34, 8, p.883.

13.   NF.A., 35-50: Iron and Steel Prods., Steels for
      Hot Dip Galvanizing, June 1984 (Zinc Development
      Association, London).

14.   Information   Feuerverzinken   No.   4,   Beratung
      Feuerverzinken,   Dusseldorf   (Zinc   Development
      Association, London).

15.   J.J. Sebisty and G.E. Ruddle, Ibid., reference 3.

16.   V. Leroy, C. Emond, P. Cosse and L. Habraken,
      Ibid., reference 3.

17.   A.N. Kirkbridge and A. Wells, in Procs. of 15th
      International Galvanizing Conf., Rome, 1988, Zinc
      Development Association, London, p.GC1/1.

18.   D.H. Rowland, Trans. Amer. Soc. Metals, 1948, 40,
      p.983.

19.   British Standards Institution, B.S.I. Standard
      Method of Test for Metallic and Related Coatings,
      B.S. 5411: Part 5: 197.

20.   R.M. Hudson, H.E. Biber, E.J. Oles and C.J.
      Waring, Met. Trans. A., Dec. 1976, 7A, p.1857.

21.   R.W. Richards and H.E. Clarke, International Lead
      Zinc Research Organization, Project ZM-375, Report
      March 1992, ILZRO, Research Triangle Park, North
      Carolina, U.S.A.

22.   H.J. Böttcher, Galvanizing of Silicon Containing
      Steels, Report Presented to the European Advisory
      Committee on General Galvanizing, Zinc Development
      Association, London, Feb. 1989.

1.5.7
Adhesion of Electroless Nickel Coatings on Fibre Composite Substrates

H. Buchkremer-Hermanns,[1] G. Matheis,[1] H. Weiß,[1] and U. Fischer[2]

[1] LABORATORY OF SURFACE ENGINEERING, SIEGEN UNIVERSITY, D-5900 SIEGEN, GERMANY

[2] L. BREITENBACH CO., D-5900 SIEGEN, GERMANY

1 INTRODUCTION

Carbon fibre reinforced polymers (CFRP), a special type of 'advanced composites', have been used for aerospace applications for several decades because of their combination of light weight, high strength and high stiffness.[1] Nowadays there are various additional fields for application of these materials. For example, they are tried for many highly stressed components in general engineering in order to reduce the weight and thereby the inertia of moving parts.[2] This latter use, however, often requires wear resistant surfaces, which are not provided by this material itself. For this reason, CFRP rolls for textile and printing industry should be fitted with a metal coating for tribological purposes. Some important techniques for metallizing fibre composite materials are the following:[3]

- Thermal Spraying
- PVD Processes
- Electroless/Electrolytic Plating.

The most important commercial process for metallizing plastics is electroless plating. The metal deposited is perfectly uniform in thickness over the whole substrate surface. Therefore, parts of complicated and complex shape can easily be metallized. The electroless metal deposit has two main functions: one is to provide an electrically conductive layer, which subsequently can be electrolytically plated like other metals, and second to provide sufficient adhesion to anchor the electroplate to the substrate.

In this paper we report upon investigations on the metallization of cylindrical CFRP rolls by electroless nickel as a first layer of a more complex wear resistant coating system. In particular, the influence of substrate surface preparations on the adhesion of the metal was studied by some selected testing methods and by means of scanning electron microscopy. We focus on discussions of the complex adhesion mechanisms responsible for this particular substrate/coating system.

2 EXPERIMENTAL

Substrate Material

Cylindrical hollow rolls consisting of a carbon fibre reinforced epoxy resin were utilized as substrate material. The volume portion of the continuous
high strength carbon fibres amounted to ca. 60 %. The rolls were manufactured by the wet winding method.[4] Due to this fabrication technique the rolls
exhibited a closed polymer matrix coating of nearly 50 μm thickness at their
surface. In order to limit the sample size the original rolls of 140 mm diameter
were cut into spherical segments of ca. 20 mm width and ca. 50 mm length.

Substrate Treatments

Electroless plating on non-conductive materials requires a preparation
of the substrate surface in order to eliminate contaminations and to generate
a proper surface micro-roughness. The surface treatment is a critical process
step, because it crucially influences the formation of an adherent continuous
electroless coating. The preparation of the CFRP sample surfaces was done by
mechanical treatment procedures and by etching with acidic solutions containing chromium.

The mechanical roughening was carried out by grinding the sample surfaces with wet SiC - abrasive paper of coarseness 400, 800 and 1200. In the
latter case a subsequent polishing with an alumina/water suspension was performed. Alternatively, blasting with SiC - powder (80 μm) was used to mechanically attack the surface. The best blasting parameters turned out to be, 100
mm sample distance, 0.6 N/mm^2 blasting pressure.

For chemical pickling of the CFRP samples, acids were selected, which
are normally used for ABS plastic etching. One etchant formulation (type 1)
consists of CrO_3 and H_2SO_4. Another formulation (type 2) was a nearly saturated aqueous solution of chromium trioxide. In order to prevent deeper damage
of the carbon fibre composite substrates, a short exposition time of three minutes at a temperature of 63 $\pm$ 3 °C was chosen. For removal of hexavalent
chromium compounds, which poison the catalyst in the next process step, the
samples were dipped into a solution of $FeSO_4$ in sulfuric acid to reduce Cr(VI)
to Cr(III).

Metallizing

Following this treatment a thin adherent conductive layer was deposited
by means of a multistage electroless plating procedure. The chemical metallizing was performed by using bath series of Max Schlötter Co., Germany.[5] For
activation a so-called one step catalyst - a stabilized Pd/Sn colloid - was used.
Nickel was applied as coating material, because it belongs to the kind of me-

tals which function as catalysts during auto-catalytic growth of the electroless layer and is able to minimize stresses at the interface substrate/deposit due to its ductility. The acidic nickel bath operates at elevated temperatures. It contains a phosphorus compound as the reducing agent and it deposits a nickel alloy with small portions of phosphorus. Subsequent to activating and removal (conditioner) of disturbing tin compounds the samples were intensively rinsed with deionized water.

For thermal shock loading experiments the thin conductive Ni-layer was electroplated by means of a copper sulfate bath to obtain thicker coatings. Copper deposition was done for 5 minutes at a current density of 0.3 A/dm^2 and for 15 minutes at a current density of 1.0 A/dm^2.

Surface Roughness

Measurements of the surface roughness of the pretreated CFRP samples were carried out (DIN 4768 part 1) by the profile method using a 'Perthometer C 5 D' of Mahr - Perthen Co. The parameter R_z (mean peak-to-valley height) was determined in order to investigate the influence of the surface roughness of the different pretreated substrates on the adhesion of the metal deposit.

Adhesion Testing Methods

Quantitative and qualitative testing methods have been used to estimate the adhesive strength of the substrate/metal system.

Adhesive Strength in Tension. For quantitative determination of the adhesion of the metal deposit on the CFRP substrates the pull off test (DIN 50160, part B) was modified by cementing two curved sample holders axially to the cylindrical samples (Figure 1). Therefore, realistic stress conditions during the test can be anticipated. The adhesion strength values result from the maximum force at the moment of fracture divided by the real, pulled off coating area. One has to take into consideration, that due to the curvature of the specimens a combined tensile and tangential stress at the interface substrate/coating arises as a function of the angle between the applied force and the surface normal. For this reason, the fracture stresses determined for the curved specimens can only be compared with those of plane samples in a limited way.

Thermal Shock Loading. An important qualitative test for estimation of the adhesion between nickel deposit and the substrate, thermal shock loading of the samples (elctrolytic copper (10 - 15 μm), 10 cycles, 100 °C for 10 minutes and -196 °C for 1 minute) was performed. The extreme temperature difference of 296 °C induced stresses in the composite metal/CFRP due to differences in the elastic modulus and the thermal coefficients of expansion of the coating material and the substrate. If the metal did not strip off, the composite quality of the tested system was not affected by thermal loading and a sufficient adhesion could be expected.

Scratch Test. The scratch test (diamond-stylus, constant load, scratching at right angle to carbon fibre direction) was also used to get - rather

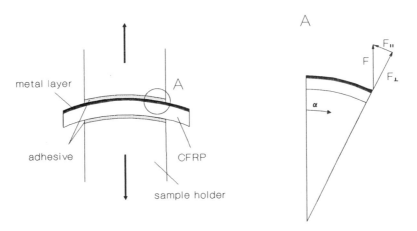

Figure 1 Pull off test of the curved CFRP samples

indirect - information about the adhesion of the electroless metal coating. For this reason the furrows and their edge zones were analyzed by means of electron microscopy.

SEM and EDX Examinations

A scanning electron microscope (SEM), type CS 2 -91, CamScan Co., was used to analyse the surface topography of the pretreated and metallized samples and the appearance of the sample surfaces after the adhesion tests. In addition, SEM was employed to estimate the thickness of the nickel deposit. In order to avoid charging of the polymeric samples a thin gold film was sputtered onto the sample surfaces. Qualitative determinations of the chemical composition of several sample surfaces were done by energy dispersive X-ray micro analysis (EDX), type 9800 - ECON IV, EDAX Co.

3 RESULTS AND DISCUSSION

The various pretreatment procedures result in different characteristic surface structures of the CFRP samples.

SiC - blasting generates an extremely nonuniform surface full of fissures (Figure 2). The surface consists of a large number of dimples and furrows of variable size. This crater-like surface topography is documented by the maximum R_z values (Figure 3a) of the preconditioned samples. In contrast to grinding no deeper abrasion of the outer epoxy resin layer occured, so that carbon fibres were left completely embedded in the matrix material.

Etching (type 1) of the blasted specimens entails a small reduction of the surface roughness. Clefts and pores of submicroscopic dimensions, spread over the whole sample surface, are formed by the chemical attack (Figure 4). As a first step, the chromic-sulfuric acid seems to cause swelling of the matrix surface.

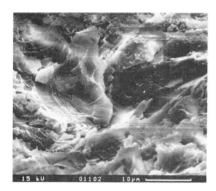

Figure 2 CFRP surface topography after blasting

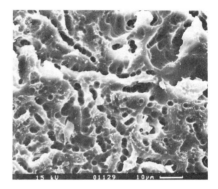

Figure 4 CFRP surface topography after blasting + etch 1

If grinding was performed, the outer matrix layer of the CFRP samples was removed appreciably, while carbon fibres became exposed to some extent. The degree of exposure was determined by the grain size of the abrasive. The fibres are only partially uncovered by grinding with the finest grain (1200) and subsequent polishing. They are either still enveloped or considerably embedded in the polished matrix. Using abrasive paper with larger grain sizes (800, 400), a similar surface topography is generated (Figure 5). Nevertheless,

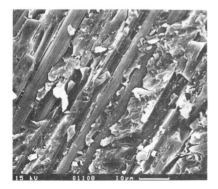

Figure 5 CFRP surface topography after grinding (800)

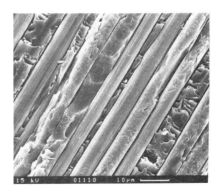

Figure 6 CFRP surface topography after grinding (800) + etch 1

a coarse-grained fractured surface can be detected, and a larger exposure of fibres is seen. It is obvious that the grain size of the abrasive corresponds directly to the surface roughness of the ground CFRP specimens (Figure 3a).

The combination of grinding (800) and etching (type 1) reveals that the additional, chemical attack on the samples causes extensive removal of the epoxy resin matrix in the zones near the surface, whereas the carbon fibres are inert to the acidic solution (Figure 6). Yet, at areas where parallel fibre ropes do not meet, the matrix largely remains and shows a micro-roughened surface structure. In case of etching with CrO_3 in H_2O (type 2), the chemical attack on the epoxy resin causes a nearly complete exposure of the outermost fibre layer. In any case, etching generates increased roughness values.

The different pretreatment procedures serve to roughen the specimen surfaces. An increase of the surface roughness, which is connected to an enlargement of the real surface, should finally also result in an improvement of the adhesion between the composite material and the metal, because more van der Waals binding forces can be effective on one hand and a better mechanical anchoring should be possible on the other hand. The comparison of pull off test results (Figure 3b) with the experimental surface roughness values (Figure 3a) seems to demonstrate the correlation of the surface roughness with the adhesive strength. Nevertheless, the highest adhesion strength was measured with a sample of medium roughness. An enlargement of the surface is an important prerequisite for good adhesion, but especially mechanical interlocking demands a suitable surface topography of the samples. The combination of grinding and etching (type 1) creates channel-like cavities between parallel carbon fibres. Additionally, the residual outer matrix has a micro-roughened surface (fissures, dimples) similar to the surface of the blasted samples. Metallization by means of the electroless nickel bath took place in these two regions and the anchoring of the deposit in both of them seems to be responsible for the best adhesion strength values.

The exposure of the carbon fibres to a greater extent (etching bath type 2) results in a lowering of the experimentally determined adhesive strength, because the carbon fibres are only weakly bonded to the surrounded epoxy resin matrix, although they are strongly connected to the enveloping nickel layer. A SEM photgraph (Figure 7) of the pulled off back side of the nickel deposit confirms this weakening. The chemically deposited metal layer is seen embracing a torn out carbon fibre.

Therefore, only a certain degree of exposuring carbon fibres should be allowed for an optimum adhesive strength of the complex C-fibre/matrix/Ni-deposit composite system. Based on the interpretation of several SEM photographs we anticipate a mechanical adhesion mechanism. The metal deposition takes place in the above mentioned cavities and fissures, in which the metal 'claws' reside. Nevertheless, we are not able to exclude completely adhesion by chemical binding forces, because the chemical pickling procedure could generate reactive (polar) groups at the substrate surface, which make an interaction to the metal possible via chemical bonds.

The surface topography of the blasted samples sharply contrasts to that

a)

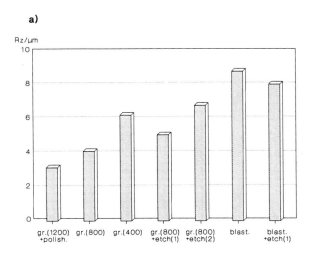

b)

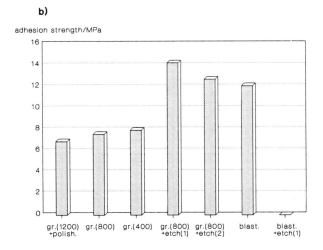

<u>Figure 3</u> a) R_z values of the different pretreated CFRP surfaces b) Adhesion strength values determined by the pull off test

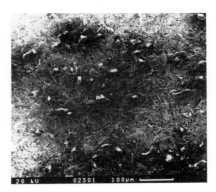

Figure 7 Deposit back side after pull
off test of a ground/etched(2) sample

Figure 8 Deposit back side after pull
off test of a blasted sample

of the ground/etched samples. At the exterior surface of the matrix a characteristic interlocking roughness is present, whereas the carbon fibres are completely embedded in the matrix. In the course of our studies on electroless
copper coatings on carbon fibre reinforced polymers we showed that SiC -
blasting generates cavities with undercuts, which are fundamental for mechanical anchoring. Adhesion strength values up to 18 MPa could be obtained.[6]
The adhesion strength of electroless nickel determined in the present study is
somewhat lower (30 %). We assume that, analogous to our observations for
electroless copper plates, mechanical anchoring is the crucial adhesion mechanism, the principle of which corresponds to a 'dovetail effect', discussed for
electroless coatings onto ABS plastics.[7] We have done EDX and SEM examinations of the deposit back side of metallized blasted specimens subjected to
the pull off test. The excellent adhesive strength of the metal/matrix system
is reflected in the detection of many epoxy resin particles (bright spots with a
dark border) on the pulled off metal layer (Figure 8). A separation between the
neighbouring phases does not seem to eventuate but rather a cohesion fracture
in the outermost matrix layer. The occurence of these fractures in the polymer
material is also described in literature for ABS and polypropylene plastics.[8,9]
Therefore, the adhesive strength of the substrate/nickel system is limited to
the tensile strength of the epoxy resin, the cross-section of which could be weakened additionally by uncompletely filled cavities.

The highest adhesion strength values are found for ground/etched or
blasted CFRP sample surfaces. Qualitative scratch test studies confirm these
results. The edge zones of the scratch furrows of samples with good adhesion
exhibit only small pieces of peeled off or lifted nickel metal (Figure 9). The
metal layer is almost undamaged up to the furrow edges.

Figure 9 gives an impression of the surface structure of an electroless
nickel layer. The thin deposit (ca. 0.7 μm) adapts the contours and consequently the roughness of the prepared surface.

Figure 9 Scratch furrow of a metallized sample (grain size 800, etch 1)

By means of EDX, phosphorus (ca. 8 %) as alloying constituent of the different metallized sample surfaces could be analysed.

The behaviour of the coated CFRP samples at definite temperature changes is a measure of the adhesion quality. Thermal shock loading of all metallized specimens did not lead to any lifting off of the metal layer. In addition, a tape test, which was performed by pulling off an adhesive tape strip at right angle to the test surface did not reveal any aggravating adhesion lost after thermal stress.

Despite several attempts to metallize the blasted and etched samples we have not been successful in producing closed nickel coatings on the sample surfaces. Partially coated zones were sharply separated from uncoated zones. The reason for this behaviour could be due to uncomplete removal of the catalyst poison (Cr(VI) compounds) from the small pores generated by chemical pickling.

4 SUMMARY AND CONCLUDING REMARKS

Our investigations show that it is possible to metallize cylindrical carbon fibre reinforced epoxy components by a thin continuous electroless nickel film, containing small amounts of phosphorus. The adhesion strength of the metal deposit can be controlled by variation of the substrate treatment techniques (blasting, grinding, etching). In order to determine the adhesion between substrate and metal deposit, quantitative as well as qualitative testing methods were utilized. The best adhesive strength values with normal fracture stresses of about 14 MPa were achieved by combination of mechanical (grinding) and chemical 'surface roughening' processes. The values are substantially higher than those described in the literature.[10] The authors suggest mechanical anchoring as the decisive adhesion mechanism in the substrate/coating system investigated.

The CFRP samples were coated by an electrically conductive nickel layer in order to prepare them for deposition of wear resistant metal layer(s) by means of electrolytic plating. In order to reduce the roughness of the chemically metallized surfaces, galvanic baths with strong smoothening characteristics are recommended for this treatment.

REFERENCES

1. D.F. Adams, 'Carbon Fibres and Their Composites', ed. E. Fitzer, Springer-Verlag, Berlin, 1985, Chapter 5, p. 205.
2. I.L. Kalnin, 'Carbon Fibres and Their Composites', ed. E. Fitzer, Springer-Verlag, Berlin, 1985, Chapter 6, p. 246.
3. F.J. Gammel and R. Suchentrunk, 'Metallische Auflagen auf faserverstärkten Kunststoffen', MBB-Z-0229-88-PUB=Acc-Nr.3544, 1988.
4. R.W. Lang, H. Stutz, M. Heym and D.Nissen, Angew. Makromol. Chem., 1986, 146/146, 267.
5. Directions for use, Dr.-Ing. Max Schlötter GmbH & Co.KG, Geislingen, Germany.
6. H. Buchkremer-Hermanns, M. Menningen, and H. Weiß, to be published.
7. W. Riedel, Galvanotechnik, 1966, 57, 579.
8. H.P. Sasse, Oberfläche, 1970, 5, 27.
9. H.P. Sasse, 'Kunststoff-Metallisierung', ed. R. Suchentrunk, Eugen G. Leuze Verlag, Saulgau/Württ., 1991, Chapter 4, p. 137.
10. R. Toupser, H. Brüntrup, H. Heckl, 'Study of the Application of Carbon Fibre Epoxy (CFE) Technology to Microwave Filters', Siemens AG, Nov. 1976.

Section 1.6 Ceramic and Glass Ceramic Coatings

1.6.1
Design of Glass–Ceramic Coatings

D. Holland

CENTRE FOR ADVANCED MATERIALS, PHYSICS DEPARTMENT,
WARWICK UNIVERSITY, COVENTRY CV4 7AL, UK

1 INTRODUCTION

The general purpose of a coating is to change the surface properties of a material, in order that the beneficial bulk properties can be exploited under otherwise untenable conditions. The most obvious need arises where corrosion or erosion of the substrate material would otherwise occur but there are many other instances of incompatibility. These include the need to provide electrical isolation or a low permittivity support for conductors; heat dissipation from active devices; and graded thermal expansion change between the substrate and some surface mounted component. In other instances, it may be the coating itself which is the active component and the bulk material is a means of providing mechanical support, cooling, electrical integrity etc. Whatever the purpose of the coating, the criteria for design and fabrication of a successful coating are the same. Firstly, the coating must have the desired physical and chemical (and sometimes aesthetic) characteristics and secondly the coating must be adherent to the substrate material. This latter requirement usually depends on two factors, close matching of thermal expansion coefficients (TEC) of the two materials and formation of a strong bond between them. Provision of all the desired properties in one material means that careful tailoring of that material must be possible. Glass-ceramics are a class of materials which provide this capability.

Glass-ceramics

Glass-ceramics are formed by the controlled crystallisation of glasses and their properties P can generally be described as the sum of the properties P_i of the crystal phases produced, modified by the volume fraction V_i of each crystal phase present

$$P = \sum_i P_i V_i \qquad (1)$$

Since chemical composition is continuously variable within the glass forming range of a particular system, then there is potential for fine-tuning of the phase composition and microstructures of glass-ceramics to give the desired value of a particular physical property. The problems lie in satisfying all of the requirements simultaneously and in then producing a viable coating. Figure 1 for example shows the TEC values for the crystal phases which can be formed in the $MgO-Al_2O_3-SiO_2$ system. If a coating is required which will be compatible with silicon (TEC = 3 ppm/K) then the combinations of phases which will give this value are limited. Mullite would appear to be an obvious choice but its refractory nature means that it would be difficult to process. Cordierite has a TEC which is too low but it can be combined with the correct proportion of a higher TEC phase to give a glass-ceramic with the correct TEC. Again, to keep the processing temperatures to a reasonable level, phases such as enstatite are preferable to spinel. Once the desired crystal phase content has been identified, then the chemical composition and its melting characteristics can be calculated from the phase diagram.

2 THERMAL EXPANSION MATCHING

If two materials, of differing thermal expansion coefficients (TEC) are bonded together at some temperature T and then cooled to temperature T_0, then the stress generated as a result of differential contraction is given approximately by equation (2).

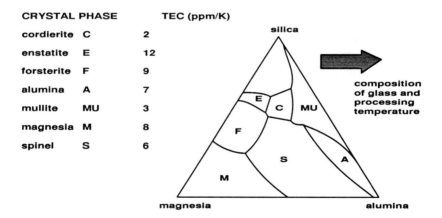

CRYSTAL PHASE		TEC (ppm/K)
cordierite	C	2
enstatite	E	12
forsterite	F	9
alumina	A	7
mullite	MU	3
magnesia	M	8
spinel	S	6

Figure 1 Illustration of the design of glass-ceramic composition to give the desired value of one physical parameter-namely thermal expansion coefficient

$$\sigma = \frac{E(\alpha_2 - \alpha_1)(T - T_0)}{(1 - \nu)} \tag{2}$$

Where α = TEC, E is Young's modulus and ν is Poisson ratio. The Young's moduli and Poisson ratios of the two materials should be similar for successful coating. If σ exceeds the breaking strength of the coating or of the interface, then cracks may occur within the coating or it may delaminate from the substrate. Even if failure does not occur, bowing of thin-section, coated substrates can arise. This can create problems if there is to be subsequent processing of the material which requires a flat surface, as in the screen printing of components onto electronic substrates.

3 ADHESION

Interlayer formation

The nature of the bond between coating and substrate can include a mechanical "keying" element but a strong bond requires there to be chemical interaction between the two materials. This should be sufficient to provide continuity of bonding. If no such continuity exists, then there are only weak van der Waals forces to provide adhesion (Figure 2). Chemical continuity can take the form of a monolayer of interaction product or a more extensive layer can be formed. If we consider specifically coatings on metals, this means that the interlayer formed is in most cases the metal oxide. This oxide can be formed by in-situ reaction with the coating during processing, or it can be preformed before the coating is applied. The choice of procedure depends on the nature of the oxide. If it is well-adherent to its parent metal and is itself

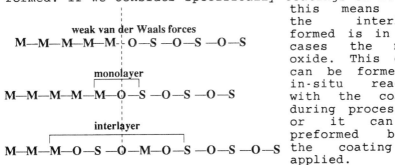

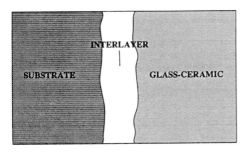

Figure 2

Interlayer formation between a metal and glass-ceramic

mechanically strong, then preoxidation is preferable, giving a layer 1 to 2 microns in thickness which is more likely to survive the coating process. If the natural oxide of the metal does not fulfil these criteria, then either special oxidation conditions must be used to produce a suitable oxide or a very thin layer of oxide must be produced by in-situ reaction. This may occur as a natural part of the coating process, or additional components must be introduced into the coating composition to induce the oxide formation by a redox process. The following examples illustrate this need for careful selection:-

a) Chrome-iron

Oxidation of a chrome-iron alloy in air generates an oxide which is largely ferric oxide, neither strong nor adherent. A chromium oxide layer is both strong and adherent but requires the oxidation process to be carried out in a H_2O/H_2 atmosphere.

b) Molybdenum

Molybdenum oxides are not only weak but volatile and therefore preoxidation to give a 1 to 2 micron layer is out of the question. In this case we have to rely on the formation of the oxide during coating. This has important consequences for the firing of the coating. Obviously the coated assembly cannot be fired under normal atmospheric conditions, since any exposed areas of molybdenum (and this may include the interface with the coating during the early stages of sintering) will be subject to rapid oxidation. An inert atmosphere, such as nitrogen will prevent atmospheric oxidation of the molybdenum but the interlayer will then have to form by a redox reaction involving the coating. Molybdenum is not a highly reducing metal and will not reduce for example magnesium aluminosilicate. In this case the firing atmosphere has to be carefully adjusted to provide sufficient oxygen to produce the desired interfacial oxide of only a few atomic thicknesses. This is far from simple, since oxygen is generally also required to remove, by oxidation, the organic binders used in the coating process. In practice the oxygen concentration (usually <10ppm) is achieved by a trial and error process and is dependent on the loading in the furnace and coating thickness.

c) Titanium

Again, the oxide produced on titanium by processing in air is weak and poorly adherent. Processing in nitrogen leads to formation of nitrides which are not immediately compatible with a silicate coating. Titanium is a highly reducing metal which reacts with silicates to produce silicides and oxygen gas, leading to foaming of the fluid glass coating. In this instance, the interfacial oxide has to be produced by controlled redox reaction with the coating. This is achieved by addition of a so-called "adherence oxide" to the coating composition. For instance, CoO, at a concentration level of approximately 1% will produce a reaction as follows:-

$$Ti + 2CoO \cdots\cdots> TiO_2 + 2Co \qquad (3)$$

The thin TiO_2 layer is then a barrier to further reaction between the titanium and the silicate [1]. The firing process has to be carried out under argon to prevent oxidation and nitridation. Once formed, the glass-ceramic coating is a barrier to oxygen ingress, allowing use of the metal at elevated temperatures.

Interlayer stability

Reaction at the interface should not be so extensive that it changes the composition and hence behaviour of the coating. Pask[2] reasoned that the important characteristic of the chemical interaction is the production of a stable interlayer compound. The stability of this interlayer depends on the presence of an interfacial region which is saturated with reaction product. Saturation prevents further reaction between the coating and the substrate/interlayer which would lead to either depletion of the interlayer and loss of adhesion, or to degradation of the coating itself and consequent change in properties. The chance that saturation will be achieved depends on the structural role of the ions derived from the metal substrate when they enter the glassy coating at the firing temperature. If the ions are glass intermediates (i.e. they can substitute for silicon) then dissolution is low and diffusion slow. If the ions are glass modifiers (i.e. they play no part in the glass network) then dissolution is high and diffusion fast.

a)Mild steel
The oxide formed on mild steel at the firing temperature is non-stoichiometric FeO. The divalent Fe^{2+} ions from this oxide are glass modifiers and therefore dissolve and diffuse rapidly to give high concentrations of iron up to about 40 microns into the coating. Figure 3 shows a micrograph of a cross-section through the coated metal [3]. The metal is to the right of the micrograph. At the interface, there are a few residual pockets of FeO which have not dissolved in the coating. Growing from the Fe/FeO interface is a lithium zinc iron silicate solid solution phase $[L(ZF)S_{ss}]$. As we progress further into the coating, the high Fe^{2+} content suppresses the crystallisation of the normal phase composition of the glass ceramic and $L(ZF)S_{ss}$ crystals are seen. These are gradually replaced by quartz and lithium silicate crystals as the concentration of Fe^{2+} decreases. Figure 3 shows that this results in a zoned structure throughout the coating. The elemental profiles can be measured by EDX

throughout the coating and individual glasses prepared
corresponding to different positions along the
profiles. Devitrification of these glasses show similar
phase composition to those observed in the coatings
(Figure 5) and the thermal expansions of these glass-
ceramics can be measured to show the TEC profile
through the coating (Figure 4) [4]. Inspection of
equation (2) shows that there will also be a stress
gradient throughout the coating - being tensile,
immediately adjacent to the metal, becoming compressive
in the 15 to 50 micron region and then becoming
slightly tensile again. Direct measurement of this
stress profile has not yet been possible.

b) Molybdenum
As stated above, a very thin layer of molybdenum oxide
(of unknown stoichiometry) can be formed. It is most
likely that the oxide contains Mo^{4+}, Mo^{5+} and Mo^{6+} in

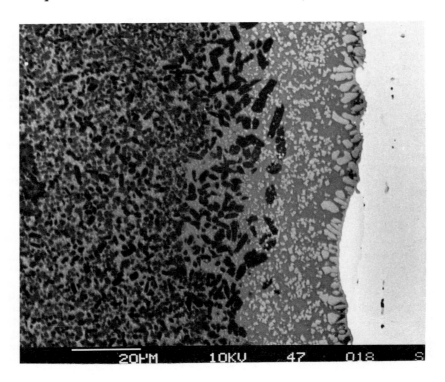

| quartz
+ lithium
silicate | lithium
silicate | lithium
zinc iron
silicate ss | FeO | mild
steel |

Figure 3 Crystallisation throughout the cross-section of a
coating on mild steel into which Fe^{2+} has diffused
from the interface

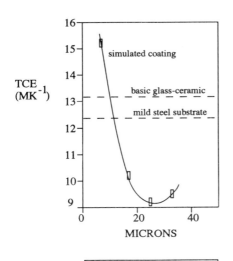

Figure 4

Thermal expansion coefficients of glass-ceramics made to simulate the composition gradient adjacent to the metal/coating interface of the lithium alumino-silicate coating on mild steel shown in Figure 3. The TCE of the bulk coating is only reached at a distance of 60 microns.

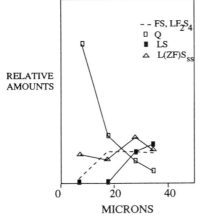

Figure 5

Crystal phase composition of the glass-ceramics made to simulate the composition gradient of the system in Figure 3. This phase composition profile is responsible for the thermal expansion profile in Figure 4.

some combination. All of these are network formers and therefore of low solubility and diffusivity. This is observed in practice since no molybdenum concentration can be detected in the coating, adjacent to the interface, by means of EDX. This is not a very sensitive technique and would not be expected to detect molybdenum at less than 0.1 or 0.2%. The absence of diffusion means that there is no modification to the coating composition and the uniform microstructure of the bulk glass-ceramic is observed (Figure 6).The presence of some dissolution and diffusion can be deduced indirectly from the dielectric behaviour of the coating at a range of frequencies. Figure 7 shows the variation in the imaginary component of permittivity as a function of frequency for different thicknesses of coating. For thick coatings, the absolute values and variation with frequency are comparable to the bulk values. However, as the coating becomes thinner, the

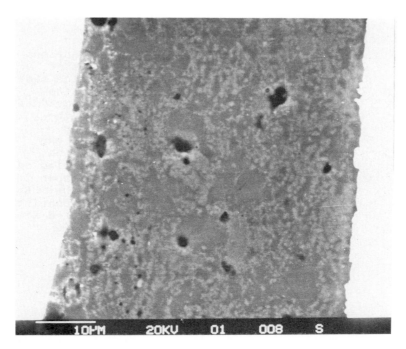

<u>Figure 6</u> Glass-ceramic coating on molybdenum showing a
uniform microstructure and no sign of diffusion
from the interface (metal is on the right)

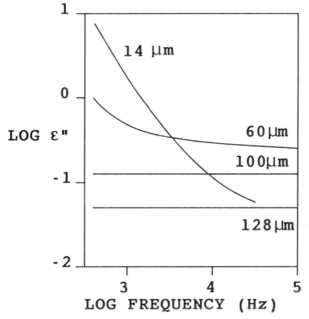

<u>Figure 7</u>

Variation of
the imaginary
component of
permittivity
with frequency
for a glass-
ceramic
coating on
molybdenum.
Measurements
are shown for
coatings of
different
thicknesses.

absolute values change and there is a strong variation with frequency. The material becomes more lossy at low frequencies, characteristic of the presence of ions which are less tightly bound in the material. It seems that this is an indication of the presence of molybdenum ions in close proximity to the interface but at low concentrations [5].

c)Copper
Figure 8 shows the cross-sectional micrograph of a lithium aluminosilicate coating on copper. The oxide formed at the interface is cuprous and the Cu^+ ion is very mobile indeed. The copper substrate is to the right of the micrograph and very bright contrast areas in the coating adjacent to the metal are remnants of the original oxide film on the metal. Similar bright areas can be seen within the body of the coating. These result from the high solubility and diffusivity of the cuprous ions at the firing temperature. On cooling the coating, the solubility limit of the cuprous ion is exceeded and the oxide precipitates out. Diffusion has taken the cuprous ion throughout the whole of the coating.

d)High T_c superconductor.
Figure 9 shows a different set of materials. This is not a conventional aluminosilicate glass-ceramic on a metal but a superconducting glass-ceramic based on a

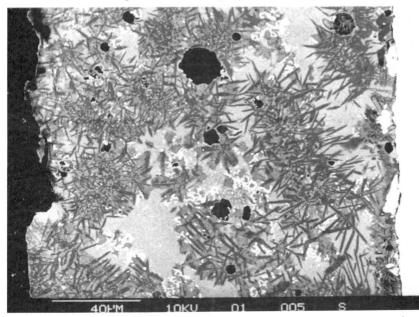

Figure 8 Lithium aluminosilicate coating on copper showing precipitation of cuprous oxide which has diffused from the interface

bismuth cuprate system coated onto an aluminium oxide substrate. The coating is from the glass forming region of the Bi_2O_3-Tl_2O-SrO-CaO-CuO system and adhesion to the substrate takes place by interdiffusion of ions to give a strontium calcium aluminate interlayer. Curiously, in the presence of thallium, this interlayer is protective against further reaction between the substrate and the coating and preserves a superconductor with a T_c in excess of 95 K. If there is no thallium, then the interdiffusion continues unchecked and subsequent degradation of the coating composition virtually eliminates its superconducting ability [6].

REFERENCES

1. F. Hong and D. Holland, <u>Surface and Coatings Tech.</u>, 1989, <u>39/40</u>, 19.
2. J.A. Pask, <u>Ceram. Bull.</u>, 1987, <u>66</u>, 1587.
3. A.J. Sturgeon, D. Holland, G. Partridge and C.A. Elyard, <u>Glass Tech.</u>, 1986, <u>27</u>, 102.
4. F. Sun and D. Holland, <u>J. Europ. Ceram. Soc.</u>, 1989, <u>5</u>, 269.
5. J.S. Thorp, M. Akhtaruzzaman, E.A. Logan and D. Holland, <u>J. Mat. Sci.</u>, 1991, <u>26</u>, 5367.
6. M.E. Yakinci and D. Holland, submitted to <u>Superconductor Sci. Tech.</u>

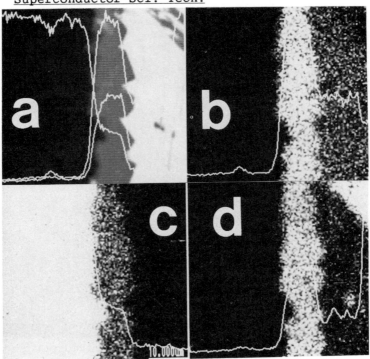

<u>Figure 9</u> Superconductor on alumina, (a)backscattered image, (b)Sr X-ray map, (c)Al X-ray map, (d)Ca X-ray map

1.6.2
Ceramic Coatings: An Ancient Art with a Bright Future

A. Ravaglioli, P. Vincenzini, and A. Krajewski

IRTEC–CNE (RESEARCH INSTITUTE FOR CERAMICS TECHNOLOGY),
FAENZA, ITALY

1 INTRODUCTION

The increasing demand for components and devices with
advanced technological characteristics, able to perform
even in extreme chemico-physical conditions, creates a
need for improving the existing techniques for materials
processing and composites elaboration on the one hand and
for devising new techniques and materials on the other.

Table 1 Main physico-chemical actions that can involve the
surface of a component in exercise.

1 Corrosion
2 Friction
3 Wear/erosion
4 Thermal action
5 Bioactive action
6 Biological action

Another specific requirement is that industry should
make use of economical, durable, and highly versatile
components.

The main actions (from the exterior) that may affect
a component during use are those summarized in Table 1.
Since most of such actions affect the surface of the
component, there is not always need for a bulk, integrally
constituted, material with remarkable technological
characteristics, except when the material is affected in
depth (a chemical attack, for example).

Table 2 Evolution of coatings on ceramics

AGE	KIND OF COATED SUPPORT	KIND OF COATING AND VARIOUS DECORATIONS
6000 B.C. Assyrians and Babylonians	fired clay	The invention of glass and first vitreous coatings.
IX-VI century B.C. Persians and Egyptians	fired clay	Ceramic paints of different colours (green, blue and yellow) and coatings based on tin. Coloured glasses.
604-562 B.C. Nabucodonor kingdom	tiles	Glasses for decorations with figures of fantastic animals as reliefs - blue cobalt, yellow and brown painted (Ischtar door-Pergamon Museum-Berlin)
521-486 B.C. Persia	tiles	Coloured glazes. An example is the frieze of the archers in Dario Ist's palace (Museum Louvre-Paris)
Ancient Greece	pots, earthenware	Objects covered with a bright layer (lustro) based on clay, alkaline elements and iron oxides (typical brown colour). Decoration with figures engraved with brown, black, red and white paints.
Etruscans	fired clay	Black ceramic (terra nigra or bucchero) obtained with clays and wood carbon in powder firing in a reducing atmosphere. The brightness was obtained by covering the objects with a very fine clay body.
XV century Italy (Gubbio, Casteldurante, Sienna, Orvieto, Pesaro, Castelli, Genova, Vicenza, Verona)and Luca Della Robbia	"terracotta"- (tiles, lunettes, architectonic, full-reliefs, etc.)	Luca Della Robbia (Firenze) begins the use of glazes for terracotta with white in the figures, green and yellow in the festoons, blue in the background.
XV and XVI century Italy (Montelupo, Deruta, Faenza, Urbino)	majolica	Many centres of Italian majolica production cover the substrate with paints, glazes and the numerous coatings with various decorations.
1368-1641	fired clay	With the most perfect techniques for the decoration over paint (also with five colours) they succeed in obtaining red glazes, flowers of peach, turquoise-blue, crimson, golden blue, black-green celadon and underglaze colours cobalt blue and copper red.
Since 1760 France-Germany	porcelain	At Paris (real manufacture of Sevrés) hard porcelain was produced (Meissen technology) at high firing temperatures with coverings and hard glazes.
Romans	fired clay	"Terra sigillata" red-coralline or red-brown, adding to the clay iron or manganese oxides with dolomite to give a degree of refractoriness during the firing.
XII century Germany (westernwald Renania)	gres	Using plastic clays sintered at high temperature, vitrified bulk was obtained with salting, through evaporation in the kiln of salt (NaCl) that condenses on the object with a red colour (vitrifying it).
900-1290 Sung dynasty	porcelain	Perfect earthenware with a very thin layer with an exceptional translucidity.
Arabian civilization	tiles	Glazes opacified by SnO_2. Black, green, yellow decorations (Utrecht)
XIII century Spain-Italy (Sicily, Puglia), Holland (Utrecht)	terracotta and majolicas	imported by Arabians and Persians. The stannifer glazes bring forth the majolica (Majorca Island). It is the Italian proto majolica period with decoration on a white background of glazes on a rough surface.
XIV and XV century Spain, etc.	majolica	Blue decoration glazes with gold gloss "lustro" (a thin metallic film appearing on the surface of the glaze). The Valencia majolica is at the top with decorations of bryionia grape vine and leaves of ivy blue and lustro coloured.

Ancient ceramists acknowledged this fact. For this reason, and because ceramic handicrafts were porous and of weak constitution (arising from the low firing temperatures), they applied coatings such as engobes and glazes. This was not purely and simply for ornamental reasons, though decorations surely played a part in attracting buyers. One of the main reasons was hygiene, in that such "sealing" of porosity could prevent the filtering through of bacteria or the fermenting of food portions (see point 6 of Table 1). Another reason may have been an attempt to reduce wear from everyday handling (point 3 of Table 1).

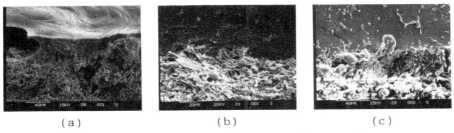

(a)   (b)   (c)

Figure 1 Examples of adhesion between glaze and:
(a) ceramic support; (b) and (c) engobes

Table 2 outlines the evolution of ceramic coating applications.

The whole skill gained over the centuries has allowed the identification of the general criteria for obtaining a good anchoring of a coating material to its support (Table 3).

Table 3 General rules to be observed by a coating to satisfactorily attach to its support

1 Coincidence of the thermal expansion coefficient values of the coating and the relative elastic moduli
2 A good wettability of the coating material at the surface of the support during its application
3 Suitable chemical affinity between the two systems

The SEM photograph in Figure 1 shows some aspects of glaze-matrix and glaze-engobe adhesion.

The criteria to obtain good adhesion can be summarized in three main rules:

1. In order to achieve a satisfactory matching of thermal expansion coefficient it is necessary either to allow the chemical composition of the coating to vary or to employ multilayers of dissimilar compositions and properties.
2. Wettability can be varied by introducing suitable additives that may affect the surface tension of the molten part during the enamelling stage and consequently also the contact angle between coating and substrate.
3. To achieve best affinity between a coating and a substrate, the former can be subjected to doping to avoid excessive ion diffusion - always during the enamelling stage - from the support into the coating itself or vice versa. Care must be taken in order to avoid or minimize the risk that mechanical properties at the interface may decay as a consequence of the changes introduced - under hot conditions - into the system, because such changes are conducive to more or less marked variations of the physical and chemical properties of the coating.

A support can itself play a rôle in determining the physical and chemical properties of a coating, especially when from the substrate towards the coating there takes place a considerable diffusion of atoms or ions originally not contained in the composition of the coating material. It is well known that one critical factor in the use of earthenware glazes is their liability to release harmful elements (e.g. lead) and that the extent of such release - being dependent on the ratios SiO_2/Pb^{2+}, SiO_2/Al_2O_3, Na_2O/K_2O, and Na_2O/B_2O_3 of the coating - makes it necessary to control the glaze support interactions in order to optimize the above ratios.

Hence the conclusion that the operations performed for optimum substrate/coating adhesion may affect the preservation of all the original properties of the material making up the coating.

2 FACTORS INFLUENCING ADHESION OF COATINGS

Additional significant factors - especially linked to surface topology - which affect adhesion are those indicated in Table 4.

Table 4 Surface-related parameters affecting adhesion of the coating to the support

```
1 Porosity

2 Roughness at the surface

3 Activated/cleaned surfaces
  (by physical or chemical treatments)
```

Porosity plays a significant rôle in ceramic glazing with respect to both the applicative technology and the adaptation of the requirements of the glaze/support. For example, our experience demonstrates that to adjust a bioglaze to an alumina support it is necessary preliminarily to look for an optimal porosity.

The so-called sandblasting process too, normally performed on metal surfaces to produce optimum roughness necessary for coating adhesion, can have a considerable influence on the end properties of the coating/substrate system. The results are strictly dependent on the rate, mass, shape, and nature of the incident particles. Especially in the case of elongated/circular grains, a number - even a large one - of them can be trapped into the surface of the material and can consequently interfere in the wettability relationship with the coating. Roundish grains are therefore to be preferred.

A decisive step is treatment of surfaces before any kind of coating is applied.

3 COATING BY VITRIFIED CERAMIC SYSTEMS

As is well known, the main field of application of ceramic glazes is that of traditional ceramics, especially earthenware, sanitary, ornamental ceramics, floor tiles, and technical porcelains.

To give an idea of the magnitude of the market for
these products it is enough to consider that worldwide the
production of floor and wall tiles alone is about one
billion sq.m. per year, equal to a road nearly three
meters wide extending from the earth to the moon. And the
glaze used yearly for such tile production is estimated at
about one million tonnes, equalling the loading capacity
of five supertankers.

Extended tests on the application of glazes on
tableware have allowed us to focus on some of the
parameters involved for controlled stability of the glassy
system.

Glazes containing oxides of coloured cations of the
first transition series were applied onto a fired ceramic
body of known composition, Table 5(a), and investigated
for lead-release, Table 5(b). The best firing temperature
was found to be 1000^0C.

The general conclusion is that the ceramic body
exerts a considerable influence over Pb^{2+} ions released
from the glazes on the basis of the different dopants.

In recent years the use of ceramic glazes has been
extended to optical devices (reflecting surfaces) and the
electronic thick-film industry. The basic glassy
composition is generally the same, while there are
variations concerning the additives.

4 COATING OF METALS

Metals are often coated to exploit all their useful bulk
characteristics while at the same time protecting their
surface from damage that might reduce or impair their
properties or the functionality of the constituent
devices.

Table 5(a) Chemical analysis of a ceramic support to be
            coated (wt %)

F.L.	SiO_2	Al_2O_3	Fe_2O_3	TiO_2	CaO	MgO	K_2O	Na_2O
12.85	48.80	14.98	5.94	0.69	10.5	2.64	1.88	0.79

Table 5(b) Pb^{2+}-release from a glaze doped with different chromophorous ions applied on a majolica support fired at 960^0C and 1000^0C and treated with 4% acetic acid for 24 hours. FB is the composition of the base glass system. (SiO_2 30.2; Pb_3O_4 17.2; ZnO 3.1; $CaCO_3$ 10.1; KNO_3 7.6; H_3BO_3 12.4; Z.Kaolin 19.4)

Chromophore and % of additions	Pb^{2+} release in p.p.m.		
	T=960°C	T=1000°C	
	Value	Value	Average
FB without Chromophore	15	1.93	2.64±0.7
	26	2.74	(25%)
	29	3.26	
FB+CuO 0.5	738	15	13.0±2.8
	170	11	(22%)
FB+Cr₂O₃ 0.5	2.45	0.33	0.46±0.2
	384	0.66	(38%)
	330	0.40	
FB+Cr₂O₃ 5.0	41	0.48	0.45±0.1
	0.64	0.30	(30%)
	664	0.56	
Fb+MnO 0.5	204	0.48	1.96±1.9
	12.7	4.20	(100%)
	206	1.21	
FB+MnO 2.5	72	7.2	11.93±9.5
	56.8	22.9	(80%)
	230	5.7	
FB+Fe₂O₃ 0.5	432.30	1.62	1.78±0.5
	80	1.4	(28%)
	236	2.31	
FB+Fe₂O₃ 3.0	2.44	3.95	6.72±4
	476	5	(58%)
	31	11.2	
FB+CoO 0.1	294	4.30	2.45±1.6
	1885	1.49	(65%)
	84	1.5	
FB+CoO 3.0	2025	4.8	28.6±20.7
	590	39.2	(72%)
	3790	41.8	
FB+NiO 0.5	53	1.61	0.82±0.7
	22.6	0.45	(84%)
	--	0.39	
FB+NiO 3.0	3.90	2.14	9.56±10.7
	7.16	21.4	(109%)
	18.3	5.13	

Such coating can be effected industrially by techniques - some of them costly - based on thermal spraying and chemical and physical deposition. These techniques are equally applicable to metals and advanced ceramics, and in some cases to plastics too.

Some parameters characteristic of such techniques, with particular reference to the typical thicknesses thus obtainable, are outlined in Table 6.

There is no doubt that of all the techniques judged suitable for the formation of thick layers, thermal methods - notably plasma spraying - are the most common in the industry both for their relatively low cost and their versatility and simplicity.

Table 6 The fundamental characteristics of different methods of deposition (by P. Fauchais and coll.)

	Plasma	C.V.D.	P.V.D.	Ionic deposition	Glazing	Electroshaping
Thickness or rate of deposition	0.01-2mm	0.1-100 μm	1-100 μm	1-1000 μm	0.5-3mm	0.5-3mm
Porosity	some %	very low	very low	very low	some %	some %
Adhesion	good	very good	poor	very good	good	poor
Material applied (coating)	any fused material	all the metals and ceramic components	idem	idem	composites or vitreous composites	ceramics
Support to be covered	many different (from metals to plastics)	metals, alloys, carbides	metals, alloys, plastics	idem	metals, alloys, ceramics	metals, alloys
Treatment temperature	< 1000°C in the bulk	600-1400°C	low	low	600-1000°C	400-1000°C
Condition for deposition	air, neuter ambient	reactive gas	10^{-5}-10^{-1} Torr	10^{-2}-10^{-1} Torr	air, neuter ambient	

Sprayable materials range from ceramics at the highest temperatures to metals and polymers obtained by "cold" plasma spraying.

It is however important to note that some of the substances of ceramic interest that are more valuable in terms of application - such as SiC, $ZrSiO_4$, and Si_3N_4 - can hardly be sprayed by thermal methods.

The plasma spraying technique has proved particularly useful to coat materials for biomedical use and notably for the application of hydroxyapatite-based phosphatic ceramics on titanium bodies, an aspect that will be discussed in detail.

5 COATING OF METAL PROSTHESES

The latest studies and technologies (e.g. plasma spraying at temperatures between 400^0C and 3000^0C) are finding a solution to problems accompanying the covering of medical prostheses - especially metallic - as a result of the differences in thermal expansion and elastic modulus between ceramics and metals. The current trend for these prostheses is the application of ceramic coatings on supports of different natures (i.e. more similar to bone mechanically) for an improved functional performance.

Early Al_2O_3 coatings applied on metallic hip prostheses were not always very satisfactory in terms of post-operative results. Today we are looking to biocompatible ceramic coatings such as hydroxyapatite for obtaining surfaces capable of self-anchoring to bone tissue to introduce the possibility of avoiding the use of surgical cements in the future.

Inert ceramics (Al_2O_3, ZrO_2, TiO_2), bioactive ceramics (hydroxyapatite, glass ceramics), and bioactive glasses can play a rôle as anti-corrosive - or at least protective - coatings of metal parts for implant use, not unlike ceramic glazes on domestic appliances. In addition, the development of new materials with bioactive properties may forecast performances of the natural-anchoring type as in the case of (a) bioactive glass systems or (b) hydroxyapatite-based compositions. This will enable cementless implantation, with undeniable advantages in terms of functionality and performance.

Application of Bioactive Glasses on Metal Alloys

The composition of bioactive glasses used either as bulk components or for coating purposes must take account of the biological effects developing on contact with living tissue. The main chemical components should therefore essentially be the ones naturally occurring in the organism.

Table 7 Typical bioglass compositions for prosthetic uses

Components	4SS5	4SS5F	4SB1SS5	KCP1	KZS3020	4SS5-N	4SS5-C	A BC-5	B BC-1	B BC-2	D BC-3	E BC-4	F	G	H	AKRAIS	RKKP	AP40
	*	*	**	**	**	**	**	+	+	+	+	+	++	++	++			
SiO_2	45	43	30	45	50	50	50	43	55	50.5	49	51	46	34	34	45	44.3	44.3
CaO	24.5	12	24.5	24.5		24.5	19.5	11	4.5	4			33	23	5	24.5	18.6	18.6
CaF_2		16						16			8	8					4.99	4.99
P_2O_5	6	6	6	6		6	6	7								6		
B_2O_3			15															
Na_2O	24.5	23	24.5			19.5	24.5	23	18.2	17	22	18.5	5	3.5	3	24.5	4.6	4.6
K_2O				24.5	30												0.19	0.19
ZnO					20													
V_2O_5									8.8	16								
Al_2O_3									3	3	6	5.5		15	23	2		
TiO_2									7.5	7								
$Ca_3(PO_4)_2$													16	12	12		24.5	24.5
Ta_2O_5														12.5	15	1	1	
Sb_2O_3															8			
CuO											1	1						
MnO_2											0.5							
CoO											0.5							
NiO																1		
Fe_2O_3																6		
Cr_2O_3																0.6		
La_2O_3																0.5	0.5	
MgO																	2.82	2.82

* L.L. Hench;
** collaborators and L.L. Hench;
+ S.K. Das, M.C. Ghose, S.S. Verma;
++ Ceravital: da H. Broemer (da 60).

In the case of bioactive glass coatings it is, however, required to adjust basic compositions to the nature of the bodies to be coated in order to ensure good anchoring and minimize the risk of contamination, e.g. as a result of a diffusion of harmful elements from the substrate towards the coating. It is for this reason that several variants are possible within the basic system (typical examples are shown in Table 7).

With metal alloys (e.g. 316L stainless steel), for the achievement of good adhesion to the metal substrate it is absolutely necessary that additives are introduced to create a thin layer of biovitreous material capable of adhesion according the well-known Dietzel's theories. SiO_2 is therefore used instead of B_2O_3, while CaF_2 replaces CaO. This lowers the softening point and decreases the thermal expansion coefficient. Oxides of transition metals can also be added to control any excess bioactivity and give better wettability in relation to the different kinds of substrate. Bonding agents used to aid adherence to the metallic base are compounds that by induced polarization can orientate the oxygen bonds to form a link to the vitreous mass.

The substances added are designed to give the specific properties indicated in Table 8.

Our experience dramatically demonstrates the enormous influence of the glass composition on the bonding strength of coatings obtained by dip-coating in molten bioactive glass. The results in Table 9 clearly indicate that mixed doping is absolutely required to reach reasonable bonding strength values.

Application of Ceramic Composition on Titanium by Plasma Spraying

For plasma-sprayed hydroxyapatite coatings an important factor is the composition and structure of the deposited layer. Commercially available materials labelled as hydroxyapatitic reveal a quite different constitution if subjected to XRD analysis. Special attention must therefore be placed on the spraying parameters needed to obtain structures approximating that of hydroxyapatite.

Table 8 The substances added to give specific properties
to the vitreous system for biomedical application
(Note: Trivalent ions tend to reduce the ability
of proteins to adhere as well as bioactivity)

Cation (added in oxide form)	Properties
Al^{3+}	Gives the glass high resistence to corrosion and has no toxic effects.
Co^{2+}	Improves the glass's capacity for bonding with the surface of metal alloys and increases resistance to oxidation. Can cause irritation.
Ni^{2+}	Stops nickel escaping from the alloy by ionic diffusion.
Fe^{3+}	Stops iron escaping from the alloy by ionic diffusion. Lowers glass melting temperature. Seems to stimulate cell growth. Can cause slight irritation.
Cr^{3+}	Stops chrome escaping from the alloy by ionic diffusion. The physiological reactions it provokes show that it has a necrotic effect and is toxic.
Sb^{3+}	Improves the flexural modulus and stabilises the glass lattice.
Cu^{2+}	Used in some cases to correct properties caused by the other cations mentioned. Does not show problems of physiological interaction in the concentrations used: clearly easily absorbed at this low concentration since it is an important component of the physiological cycle.
La^{3+}	Improves the glass's resistance to alkalis.
Ta^{5+}	Reduces the number of faults in the glass body. Stabilises the glass and make it harder. Improves the glass's wettability in relation to the metal alloy. Seems to aid cell growth.
Ti^{4+}	Gives good coating capacity but is not very soluble.

The micrographs in Figures 2-6 show various types of
porosity, grain clustering, and separation as well as the
bonds that form between a titanium substrate and the
different ceramic compounds suitable to coat titanium-
based metal prostheses, e.g. zirconium oxide, alumina,
hydroxyapatite, and bioactive vitreous systems (Table 10).

ACKNOWLEDGEMENT

The authors wish to thank Profs E. Veniale and P. Duminuco
of Pavia University (Dept. of Earth Science) for putting at
our disposal some roentgen-scanning photographs of ancient
ceramics, showing engobe/glaze interfaces.

Table 9 Maximum tensile strength of vitreous systems with multiple additions

Sample N°	Additions (wt% of base glass composition)										σ_{1mat} (MPa)
	B_2O_3	Al_2O_3	NiO	Fe_2O_3	Cr_2O_3	CuO	Co_2O_3	Ta_2O_5	Sb_2O_3	La_2O_3	
0*					0.5	0.5	0.5				9.3
1	4	2	4	3	0.5			1			<2
2	4	1	2	7	0.5			1			2.9
3	6	0.5	1	5	0.5			0.5			5.9
4	5	0.5	3	8	0.5			2			8.8
5	8	0.5	2	4	0.5			0.5			2.9
6	6	1	0.5	6	0.5			1			<2
7	6	0.5		5	0.5			2			16.7
8	5	0.5	0.5	2	0.5			0.5			0
9	6	0.5	0.5	8	0.5			0.5			9.8
10	6		0.5	4	0.5						24.5
11	3	0.25	0.25	2	0.25			0.25			17.7
12	4	1	1	2	0.25						4.9
13	0.6	1	0.4	2	0.2	0.2		0.6	0.6		6.9
14	1	2		2	0.2			0.2	0.6		<2
15		2	1	6	0.6			1		0.5	27.5
16	0.5	3	0.2	3	0.6			1	0.1	0.1	<2
17	0.6	2	0.2	2.4	0.2	0.1		0.6	0.1	0.1	33.3
18			0.2	8	0.2		0.1			0.1	27.5

* Undoped glass

Table 10 The different cases of porosity, grain clustering, separation, and the bonds that form between a titanium substrate and the various ceramics used as coatings

Ceramics for coating	Porosity	Grains agglomeration clustering	Adhesion to metallic support	Separation of the sprayed componenets	Notes
ZrO_2	Low (Figure 3)	Compact with irregular grains at variable dimensions (Figure 3)	Not optimal (see the dark border) (Figure 3)	White streakings (lamellar grains based on Fe, Ta, Ba and dark zones based on Si, Y, Ca, Fe, Mg) (Figure 3)	The separated elements by ZrO_2 grains are those acting as phase stabilizers (Figure 3)
Al_2O_3	Low (Figure 2)	Streaked and parallel	Good (Figure 2)		Zones of not perfect adhesion due to the Al_2O_3 grains penetrating the support during the sandblasting (Figure 2)
HA (hydroxyapatite)	Pitted microporosity (Figure 4)	Amorphous grains of various dimensions and shape (Figure 4)	Good (if Al_2O_3 grains caused by sandblasting treatment are not present in the metallic support (Figure 4)	Separation to a degree between Ca and P (more white zones) (Figure 5)	Cathode-luminescence image indicates the transformation in other calcium-phosphate substances (Figure 5)
Bioactive glass	High, deriving from boiling action (air, redox reactions and O_2 formation) (Figure 6)	Vitreous particles linked of semispheric shape with definite boundaries (Figure 6)	Poor adhesion to the support (Figure 6)		Fractures parallel to the contact zone within the glass at small distance from the interface

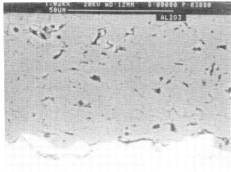

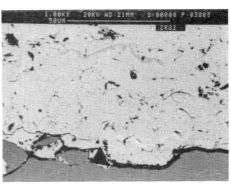

Figure 2 SEM microphotograph of interface between Al₂O₃ and titanium substrate

Figure 3 SEM microphotograph of interface between ZrO₂ and titanium substrate

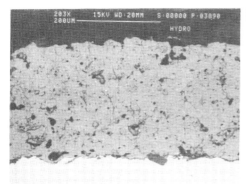

Figure 4 SEM microphotograph of interface between hydroxyapatite and titanium substrate

Figure 5 Cathode luminescence SEM microphotograph of the SEM area of Figure 4 where calcium-enriched zones (darker areas) are emphasized

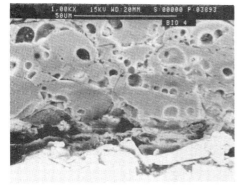

Figure 6 SEM microphotograph of interface between AkRA15 bioactive glass and titanium substrate (combination of secondary and reflected electrons)

All photos taken from ref. [22]

REFERENCES

1.   G. Tamman, "The Glassy State," Leopold Voss, Leipzig,
     1933, p.123.
2.   G.W. Morey, Jour. Amer. Ceram. Soc., 1934, 17, (11),
     315-328.
3.   A.Q. Tool and C.G. Eichlin, ibid., 1931, 14, (4),
     276-308; Bur. Stand. Jour. Research, 1931, 6, (4),
     523-552.
4.   C.G. Peters and C.H. Cragoe, Bur. Stand. Sci. Paper,
     1920, p.393, abstracted in Jour. Amer. Ceram. Soc.,
     1920, 3, (10), 852.
5.   J.T. Littleton, Ind. Eng. Chem., 1933, 25, (7), 748-
     755; Ceram. Abs., 1933, 12, (10-11), 360.
6.   O. Koerner, H. Salamang and W. Lerch, Sprechsaal,
     1932, 65, 925.
7.   E. Berger, Jour. Amer. Ceram. Soc., 1932, 15, (12),
     647-677.
8.   W.H. Zacharissen, Jour. Amer. Chem. Soc., 1932, 54,
     (10), 3841-3851; Ceram. Abs., 1933, 12, (4), 145.
9.   C.W. Parmelee, G.L. Clark and A.E. Badger, Jour.
     Soc. Glass Tech., 1929, 13, (52), 285-290; Ceram.
     Abs., 1930, 9, (5), 339.
10. E.V. Smirnova, Proceedings of 4th All Union
     Conference on the Glassy State, Leningrad, 1964, Vol.
     7, p.25, Academy of Sciences, USSR: Transl. from
     Russian, Consultants Bureau Inc., New York.
11. A. Ravaglioli and A. Krajewski, Berichte der Deut.
     Keram. Gesell., 1980, 57, (4-5), 76-79.
12. A. Ravaglioli, A. Krajewski, C.B. Azzoni and G.L. Del
     Nero, Jour. Mater. Sci., 1981, 16, 1081-1087.
13. A. Ravaglioli and G. Vecchi, Interceram, 1981, 10, 1-
     4 (Part 1) and 1982, 3, 5-6 (Part 2).
14. A. Ravaglioli and A. Krajewski, Interceram, 1984, 5,
     22-23 and 1986, 1, 20-23.
15. A. Ravaglioli, A. Krajewski, P. Ponti, R. Valmori and
     S. Contoli, Sci. of Ceramics 13, ed. P. Odier, P.
     Cabonnes and B. Cales; Proceedings Sept. 9-11 1985,
     Coll. Cl, Suppl. n.2, Feb. 1986, p. Cl 764-769.
16. A. Krajewski, A. Ravaglioli, B. Fabbri and A. Azzoni,
     Jour. Mater. Sci., 1987, 22, 1228-1234.
17. A. Piancastelli, A. Ravaglioli, A. Krajewski and F.
     Trotta, Proc. of 22nd Intern. Metallurgy Congr.
     Innovation for Quality, Bologna 17-19, May 1988, Part
     2, 1831-1840.

18. A. Krajewski and A. Ravaglioli, <u>Biomaterials</u>, 1988, 9, 449.
19. A. Krajewski, A. Ravaglioli, A. Bertoluzza, P. Monti, M.A. Battaglia, A. Pizzoferrato, A. Olmi and A. Moroni, <u>ibid.</u>, 1988, 9, 522-532.
20. D. Zaffe, A. Moroni, V. Pezzuto, S. Contoli, A. Krajewski and A. Ravaglioli, "Histological and Physio-Chemical Analyses on Transformations of some Bioactive Glasses Implanted in Long Bones of Rabbit and Sheep," Ceramics in Substitutive and Reconstructive Surgery, ed. P. Vincenzini.
21. A. Krajewski, P. Ponti, R. Valmori, M. Messori and A. Moroni, High Tech. Ceramics, ed. P. Vincenzini, Elsevier Science Publ., B.V., Amsterdam, 1987, p. 91-98.
22. A. Krajewski, A. Ravaglioli, V. Biasini, R. Martinetti, A. Piancastelli, S. Sturlese, S. Fioravanti, N. Antolotti, C. Mangano and F. Trotta, Proceedings of "Bioceramics and the Human Body," ed. A. Ravaglioli and A. Krajewski, Elsevier, London and New York, 1992, p. 236-243.

1.6.3
Ti–B–N and Hf–B–N Coatings Prepared from Multilayers and by Co-Sputtering

W. Gissler, T. Friesen, J. Haupt, and D. G. Rickerby

INSTITUTE FOR ADVANCED MATERIALS, JOINT RESEARCH CENTRE OF THE COMMISSION OF THE EUROPEAN COMMUNITIES, C.P. I, I – 21020 ISPRA (VA), ITALY

1 INTRODUCTION

Coatings of the Ti-B-N type have attracted increasing interest due to high hardness, and their high temperature and corrosion resistance. Preparation was performed by chemical vapor deposition (CVD)[1], plasma-assisted CVD[2,3], sputtering[4-6], ion beam methods[7] and ionization assisted electron beam PVD[3]. Recently two other preparation techniques were applied. The first was from multilayers of Ti-BN by subsequent annealing at moderate temperatures of about 400 °C[8]. An inter-diffusion process is thereby initiated, by which very hard phases are formed. Interestingly, these layers can be deposited almost free of macroscopic stress[9]. The second was by co-sputtering from a Ti and BN target[8]. Zr-B-N coatings were prepared recently by reactive and non-reactive sputtering[10].

In this paper a comparison of Ti-B-N and Hf-B-N films is performed. These films were obtained by the method of co-sputtering from a Ti or Hf and a BN target and from multilayers of the sequence Ti-BN or Hf-BN which were subjected to subsequent thermal treatment. The films are extraordinarily hard with values up to 65 GPa. They were examined by scanning electron microscopy (SEM), transmission electron microscopy (TEM), Auger and X-ray electron spectroscopy (AES and XPS) and glancing angle x-ray diffractometry. Hardness and Young's modulus were investigated by an ultra-low load nanoindentation method.

2 EXPERIMENTAL DETAILS

Film deposition was performed with a commercially available sputter device (MRC type 8667 A) disposing of three different sputter target stations in which targets of Ti, Hf and BN of 21 cm diameter were mounted. All targets were of 99.9 % purity. Film deposition was accomplished either by the co-sputtering or by the multilayer method[8].

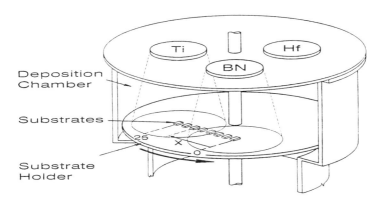

<u>Figure 1</u> Schematic diagram of substrate position in the co-
sputter deposition mode of operation

    In the co-sputtering method the metal and BN targets
were mounted in neighbouring positions and several small
substrates of dimension 1 x 2 cm were positioned along the
axis below the centre line of the two targets on the substrate table (Fig. 1). The distance of the substrate to
the BN target is given by the parameter x. As a result of
this geometry, the sputter angle and distance to the targets depend on the position of the substrate, and therefore coatings of varying thickness and composition are
obtained.

    The deposition of multilayers was performed by posi-
tioning the substrate alternately under the Ti or Hf target and the BN target. This was accomplished by rotating
the substrate holder plate. The thickness of the single
layers was controlled either by adjusting the time interval the substrate was under the targets keeping the sputter power constant or by variation of the sputter power at
the targets while continuously rotating the substrate.

    Sputtering from the Ti and Hf targets was performed
in d.c. magnetron mode and from the BN target in r.f. magnetron mode in an Ar atmosphere. A gas flow of 50 sccm was
maintained by a mass flow controller. The total pressure
was kept at $1 \cdot 10^{-3}$ mbar for multilayer preparation and at
$5 \cdot 10^{-4}$ mbar for co-sputtering. Partial pressure of the
working and residual gas was measured by a quadrupole mass
analyser, which was connected to the sputter chamber by a
pressure converter. Evacuation was accomplished by a
3000 l/s cryogenic pumping system down to 10^{-7} mbar. Prior
to film deposition the targets were sputter-etched for approximately 1 min. The sputter power at the titanium and
hafnium targets was varied between 50 and 300 W, whereas
the power at the BN target was kept constant for all experiments at a value of 1 kW. Typical deposition rates
were 15 nm/min for Ti and 45 nm/min for Hf at a sputter

power of 150 W, and 12 nm/min for BN at a sputter power of
1 kW.

Films were deposited on glass and type 316 stainless
steel. Before sputter deposition the substrates were
cleaned in trichlorethylene and isopropanol in an ultrasonic bath. The thickness of the films was usually between
1 and 3 µm and was measured by a mechanical stylus device.

As deposited films were examined by glancing angle x-
ray diffractometry using the equipment previously described[11]. The angle of incidence for most measurements
was 0.5°, giving a calculated normal penetration depth of
about 5.5 µm in boron nitride, 0.5 µm in titanium and only
0.05 µm in hafnium.

Observations of microstructure changes due to thermal
treatment were carried out using a JEOL 200 CX transmission electron microscope. The TEM specimens were prepared using an ion thinning unit consisting of two ion
guns mounted in opposing positions[12].

The composition of the coatings was investigated by
ESCA-Auger spectroscopy with a Perkin-Elmer spectrometer,
type ESCA-SAM. The elemental concentrations were determined from the areas under the Ti 2p, N 1s, B 1s and Hf 4f
peaks, respectively.

Hardness and Young's modulus were determined by an
ultra-low load depth-sensing nanoindenter from the loading
and unloading curve [13]. The loading curve was measured by
keeping the indentation rate constant (3 nm/s) and measuring the displacement until a total depth of 50 nm was
reached. A hold period of 150 s followed, to allow for a
relaxation of the induced plastic flow. Finally the unloading curve was measured, decreasing the force at a constant rate of about 40 µN/s. Elastic contributions were
determined from the unloading curve [13]. These were used
for the calculation of Young's modulus and to reduce the
loading curve to its plastic component alone. The indentation depth of 50 nm corresponds to approximately one-tenth
of the thickness of the thinnest sample. At this low penetration depth small deviations from a perfect Berkovich
diamond tip might cause errors in the absolute values of
hardness and Young's modulus. Therefore the measurements
were calibrated with a Si (111) wafer assuming a modulus
of 157 GPa independent of penetration depth. Under this
assumption also a penetration depth independent hardness
value of 12 GPa was obtained.

3 RESULTS AND DISCUSSION

Ti-BN and Hf-BN Multilayer Coatings. Fig. 2a shows a
TEM micrograph of an as-deposited multilayer of Ti-BN with
single layer thicknesses of d_{Ti} = 20 nm and d_{BN} = 8 nm,
whereas Fig. 2b shows a similar multilayer after annealing
at 400 °C for 4 h in an Ar atmosphere of 10^{-3} mbar. The

untreated sample shows a very sharp boundary between the
Ti (dark) and the BN (light) single layers. The thermally
treated coating shows that a diffusion process between the
Ti and BN layers has occurred. The Ti layers show columnar
type growth and diffusion proceeds preferentially along
the direction of the columns. Evidently, inter-penetration
of the layers was not complete in the indicated thermal
treatment time.

The observed interdiffusion process has a pronounced
effect on the hardness of the coating. Hardness reaches
extraordinarily high values well over H ≈ 50 GPa[8]. This is
considerably higher than the hardness of TiB_2 of 34.8
GPa[13]. Glancing angle x-ray diffraction investigations
suggest that multilayers of Ti-BN with d_{Ti} = 10 nm and d_{BN}
= 4 nm have predominantly fcc structure of the $TiN_x(B_y)$
type[8].

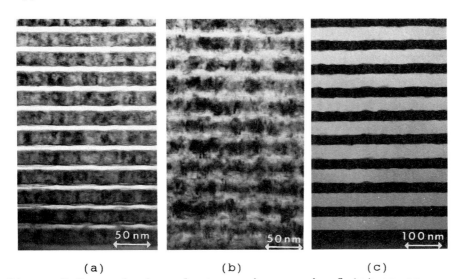

(a)   (b)   (c)

<u>Figure 2</u> Transmission electron micrograph of (a) an as-
deposited T-BN multilayer, and (b) a multilayer
which was annealed at 400° C for 4 h, and (c) a Hf-
BN multilayer which was annealed at 400°C for 70 h

This is also indicated by ESCA measurements. These
revealed a non-broadened peak of the Ti 2p line at 454.9
eV. A comparison with the values for pure Ti at 454.4 eV,
for TiB at 454.5 eV and for TiN at 455.8 eV indicates that
the Ti atom is bonded to both N and B atoms.

Due to the chemical similarity of hafnium and titanium we expect that hafnium also might form a hard phase
with boron nitride at the interface. Multilayer coatings
of Hf-BN were therefore prepared and annealed. Fig. 2c
shows a TEM micrograph of such a coating with d_{Hf} = 27 nm
and d_{BN} = 40 nm after annealing at 400° C for 70 h. Sur-
prisingly, there is no evidence of an inter-diffusion

process between the Hf and BN layers in contrast to the
behaviour of Ti-BN multilayers (Fig. 2b). The diffusion
constant of hafnium will be smaller than that of titanium
as consequence of the roughly three times greater atomic
mass of hafnium. The diffusion constant of B and N in
hafnium should be smaller than that of titanium due to the
30 % larger atomic volume. However, taking into account
that the Hf-BN multilayer coating has been annealed about
20 times longer than the Ti-BN multilayer these considerations can hardly explain the complete absence of visible
indication of Hf-BN mixing analogous to that observed in
the Ti-BN coatings (Fig. 2b). A possible explanation might
be that at the interface a thin Hf-B-N phase is formed
which acts as diffusion barrier for Hf, B and N.

     Co-Sputtered Ti-B-N and Hf-B-N Coatings. It was shown
earlier that it is possible to prepare hard Ti-B-N
coatings by the co-sputtering technique[8]. Fig. 3 shows the
hardness H and the ratio hardness over Young's modulus E
of Ti-B-N films co-sputtered at a bias voltage of - 30 V
as function of the distance x to the BN target (see Fig. 1).
Due to the inhomogeous composition along the axis between
the BN and Ti target H varies with distance to the BN target. Coatings of hardness up to 60 GPa are formed at a position $x \approx 13$cm. Fig. 4 shows the x dependence of the
boron [B], nitrogen [N] and [Ti] concentration as obtained
by ESCA measurements. At a distance x of about 13 cm to
the BN target, where H reaches its maximum value, [Ti] and
[B] are about equal. A closer inspection of the ESCA spectra shows that at x > 12 cm only one un-broadened B 1s
line at 187 eV can be observed suggesting that only one
phase, probably of the TiB_xN_y type is predominant. At x <
12 cm a second B 1s line appears at 190 eV the intensity
of which increases with decreasing x values. This indicates the appearance of another phase of the BN_y type. It
is interesting to note that also films of different composition $TiB_2(N)$ (the nitrogen content is in the percent
range) reach similar high hardness values[6].

     The ratio of H/E (Fig. 3) is considered by several
authors to characterize the material better than H and E
alone[14]. It can be expected the abrasive wear would be reduced with increasing hardness and decreasing elasticity.
This is due to the fact that, at a given hardness, large
strain can better be sustained with smaller E values.
Characteristic H/E values are 0.01 - 0.04 for metals, 0.10
for diamond and diamond-like coatings[15] and TiN[16]. The
high values observed here are therefore promising for a
good wear resistance.

     In contrast to the multilayers we would expect that
co-sputtered Hf-B-N coatings become hard, because in this
case the components are mixed during the deposition process and the hardening process does not depend on

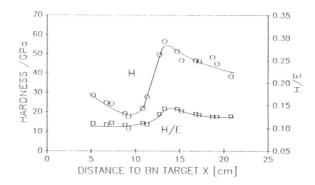

Figure 3 Hardness H and H/E ratio of co-sputtered Ti-B-N
coatings as function of the position x of the
substrate

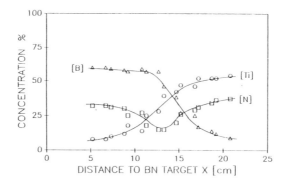

Figure 4 Relative content of titanium, boron and nitrogen as
function of the substrate position

thermal diffusion. This is actually what is observed: ul-
tra hard Hf-B-N coatings are obtained. Fig. 5 shows hardness and the ratio H/E of co-sputtered Hf-B-N films at a
bias voltage of - 30 V as function of the distance to the
BN target. A similar tendency is observed as for Ti-B-N
coatings, with highest values of both H and H/E observed
where the concentration ratio of the coating constituents
Hf/B is approximately 1 (Fig. 6). Due to the different
sputter coefficients (about a factor of 3 larger for Hf
than for Ti) the geometrical position where the deposition
rates of the metal components are nearly equal is shifted
in the case of the Hf-B-N coatings towards a smaller values of x ($\approx$ 10 cm). From ESCA line analysis - performed
analogously as for the co-sputtered Ti-BN films - it is
observed that for x < 8 cm boron is present as BN and as
HfB_xN_y, whereas for x > 8 cm only the latter compound is
present.

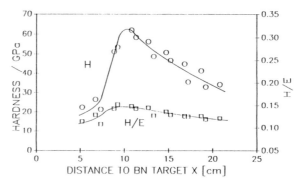

Figure 5 (a) Hardness H and H/E ratio of co-sputtered Hf-B-N coatings as function of the substrate position

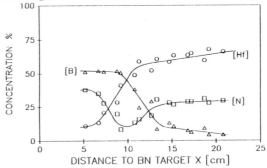

Figure 6 Relative content of hafnium, boron and nitrogen as function of the substrate position

The ratio H/E also indicates this; much smaller values of the order of 0.07 are observed near the Hf target indicating metallic behaviour. Maximum values of nearly 0.15 are observed, which suggest that the Ti-B-N films as well as Hf-B-N films should have a good wear resistance.

Fig. 7 shows two x-ray diffraction patterns obtained from a Hf-B-N film at a bias voltage of - 30 V at positions x = 9.5 and 10.5 cm, respectively. Although both samples are closely positioned their concentration ratio [B]/[Hf] is quite different with values of 1.4 and 0.8, respectively (see Fig. 6). Nevertheless both patterns display essentially the same diffraction lines being compatible with a HfB_2 structure[18] at least as concerns the position of the Hf atoms (the scattering cross section of the boron atoms is by a factor of 200 smaller). There is only a slight difference for the sample with the higher boron concentration due to an additional peak at $2\theta \approx 52°$ being also characteristic for the HfB_2 structure. The formation of a HfB_2 structure is difficult to understand at a low [B]/[Hf] ratio. The relatively broad diffraction lines, however, indicate the presence of very small crystal grains (order of 50 Å), where surface effects may play a dominant role.

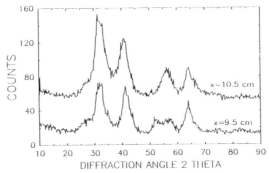

<u>Figure 7</u> Glancing angle X-ray diffraction trace of co-
sputtered Hf-B-N films deposited at a bias voltage
of -30V at positions x of 9.5 and 10.5 cm,
respectively

4. CONCLUSIONS

Very hard Ti-B-N coatings can be prepared from multilayer coatings of Ti-BN by thermal treatment at 400 °C.
Hardening occurs by a diffusion activated mixing process
between the Ti and BN layers with consequent phase transformation. An analogous preparation procedure for Hf-B-N
coatings from multilayers turned out not to be possible;
even extended thermal treatment did not result in mixing
of the interlayers. It is supposed that this is due to the
formation of a diffusion barrier at the interface. However, applying the co-sputtering technique very hard
Hf-B-N and Ti-B-N coatings up to 65 GPa were obtained. In
co-sputtering the components are mixed during deposition
and therefore the hardening process does not depend on
thermal diffusion. Ti-B-N and Hf-B-N films show similar
properties. The highest hardness and elasticity is observed if the composition ratio B/Ti and B/Hf is about 1.
In this case a hexagonal $HfB_{2-x}(N_y)$ phase is indicated by
glancing angle x-ray diffractometry. For Ti-B-N and
Hf-B-N films a high wear resistance can be expected due to
the high measured H/E ratio of 0.15.

ACKNOWLEDGEMENTS

The authors wish to express their thanks to Dr. A.
Manara and Dr. P.N. Gibson for helpful discussions in in-
terpreting XPS and XRD measurements and Messrs A. Hoffmann
and L.G. Mammarella for helpful technical assistance.

REFERENCES

1. J.L. Peytavi, A. Lebugle, G. Montel and H.
   Pastor, 'High Temperatures - High Pressures', Vol.
   10 (1978)
2. H. Karner, J. Laimer, H. Störi and P. Rödhammer,
   '12th Plansee Seminar 1989', Vol. 3, Tirol
   (Austria), 8 - 12 May 1989

3. J. Aromaa, H. Ronkainen, A. Mahiout, S.-P. Hannula. A. Leyland, A. Matthews, B. Matthes and E. Broszeit, Materials Science and Engineering, 1991, A140, 722

4. C. Mitterer, M. Rauter and P. Rödhammer, Surface and Coatings Technology, 1990, 41, 351

5. W. Herr, B. Matthes, E. Broszeit and K.H. Kloss, Materials Science and Engineering, 1991, A140, 616

6. O. Knotek, R. Breidenbach, F. Jungblut and F. Löffler, Surface and Coatings Technology, 1990, 43/44, 107

7. G. Dearnaly and A.T. Peacock, UK Patent GB 2 197 346, A and B (1988)

8. T. Friesen, J. Haupt, W. Gissler, A. Barna and P.B. Barna, Surface and Coatings Technology, 1991, 48, 169

9. T. Friesen, J. Haupt, W. Gissler, A. Barna and P.B. Barna, Vacuum, in press

10. C. Mitterer, A. Übleis and R. Ebner, Materials Science and Engineering, 1991, A140, 670

11. R.C. Buschert, P.N. Gibson, W. Gissler, J. Haupt and T.A. Crabb, Colloque de Physique, 1989, C7, 169

12. A. Barna, P.B. Barna and A. Zalar, Vacuum, 40, 1990, 115

13. M.F. Doerner and W.D. Nix, J. Mat. Res. 1986, 1, 601

14. H. Frey and G. Kienel (eds.), 'Dünnschichttechnologie', VDI Verlag, Düsseldorf, 1987

15. K. Taube, Dissertation, Universität Hamburg, Hamburg, 1991

16. J. Robertson, Surface and Coatings Technology, in press

17. X. Jiang, M. Wang, K. Schmidt, E. Dunlop, J. Haupt and W. Gissler, J. Appl. Phys. 1991, 69, 3053

18. Powder Diffraction File, Joint Committee on Powder Diffraction Standards, International Center for Diffraction Data, Swarthmore, PA, 1972 - present.

19. L.E. Toth, 'Transition Metal Carbides and Nitrides', Academic Press, New York, 1971

20. A. R. West, 'Basic Solid State Chemistry', John Wiley & Sons, New York, 1988

1.6.4
The Influence of Substrate Topography on the Adhesion of Thermally Sprayed Alumina Coatings

G. Dong,[1] D. T. Gawne,[1] and B. J. Griffiths[2]

[1] DEPARTMENT OF MATERIALS TECHNOLOGY, BRUNEL, THE UNVERSITY OF WEST LONDON, UXBRIDGE, MIDDLESEX, UK

[2] DEPARTMENT OF MANUFACTURING ENGINEERING SYSTEMS, BRUNEL, THE UNIVERSITY OF WEST LONDON, UXBRIDGE, MIDDLESEX, UK

1   INTRODUCTION

It is widely accepted that the adhesion and quality of thermally sprayed coatings depend critically on the condition of the surface of the substrate and that some form of roughening prior to spraying is essential (1). The usual procedure for substrate preparation is chemical degreasing and hand-operated grit blasting often followed by air blasting to remove loose particulates. Grit blasting is carried out using different types of grit depending upon the application and economics, and includes alumina, chilled cast iron and crushed slag.

The most widespread method of assessment of the grit blasted surface is by visual inspection by an experienced operator. Visual inspection can reveal aspects of the surface topography that are generally regarded to affect coating adhesion. For example, it can detect highlights due to smooth, reflecting areas indicative of inadequate grit blasting or variations in appearance over an area indicative of non-uniform blasting. General experience suggests that the most satisfactory grit blasted surfaces have a uniform matt appearance with few highlights and an absence of structure.

Pretreatment for thermal spraying has been developed on a trial-and-error basis and an improved scientific understanding of the critical parameters will aid the development of increased reliability in thermal sprayed coatings. This paper involves plasma sprayed aluminium coatings on plain carbon steel and is aimed at investigating the effects of grit blasting conditions and surface topography on adhesion.

2.   EXPERIMENTAL PROCEDURE

The substrate used was a plain carbon steel (080M40 grade) plate of thickness 5mm, degreased with acetone before grit blasting and air blasted subsequently to remove loose particulates. Grit blasting was undertaken using pressure-operated and suction-operated machines manufactured by Guyson International (Skipton) and alumina grit supplied by Guyson. Alumina coatings of 250μm were applied under fixed conditions with a Metco plasma spray system using an MBN torch, MCN control unit and 4MP powder feed with argon-hydrogen plasma gas at 500A arc current. The adhesion of the coating to the substrate was determined by the direct pull-off technique, using an Elcometer instrument with Araldite 2005 adhesive. The surface topography was evaluated using a stylus profilometer and an AMS2065 system.

3. RESULTS AND DISCUSSION

(a) Grit blasting conditions

Figure 1 shows the effect on coating adhesion of the distance from the blast gun nozzle to the substrate surface. The adhesion increased to a maximum of 13MPa at 50-100mm and then gradually declines to 5MPa at 700mm.

Grit blasting effectively involves introducing a stationary particle into a high velocity air stream, where the speed and direction of the particle will clearly not necessarily be the same as those of the air flow. The velocity of the particle is, in fact, related to factors, including the air stream velocity, atmospheric pressure, drag coefficient, particle size and density. The particle velocity and kinetic energy immediately before impact with the substrate will have a major influence on the development of surface roughness. When a particle impinges upon the surface of the substrate, it will lose some of its kinetic energy. Most of this lost energy is transformed into deformation (elastic and plastic) and heat. The deformation gives rise to surface roughening, which clearly increases with increasing particle kinetic energy.

In broad terms, it is known qualitatively that substrate roughness benefits adhesion and so it is likely that the optimum blast distance revealed in Figure 1 relates to the time required for a grit particle to achieve maximum velocity. The subsequent decline in adhesion is attributed to the decreasing air stream velocity with increasing distance from the blast gun nozzle.

Figure 2 indicates that increasing the blast pressure from 2.6 bar to 4.0 bar almost doubles the coating adhesion. The increased pressure is expected to raise the air stream velocity and particle kinetic energy resulting in enhanced roughening and adhesion.

The number of traverses or passes of the grit blast stream over a given area of the substrate surface was found to produce a significant effect on adhesion. Figure 3 shows that under the conditions used, there is an optimum of 4 passes to achieve maximum adhesion. Overblasting leads to a substantial deterioration: the adhesion falls from 15MPa at 4 passes to 8MPa at 15 passes.

The above coated samples were prepared by grit blasting the substrate at 90 degrees (the angle between the incident grit stream

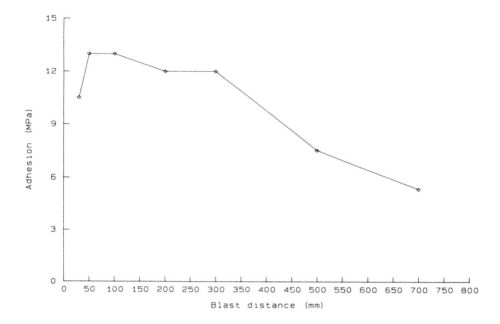

Figure 1 Effect of blast distance on coating adhesion

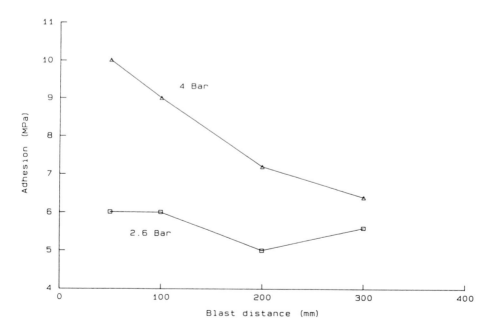

Figure 2 Effect of blast pressure on coating adhesion

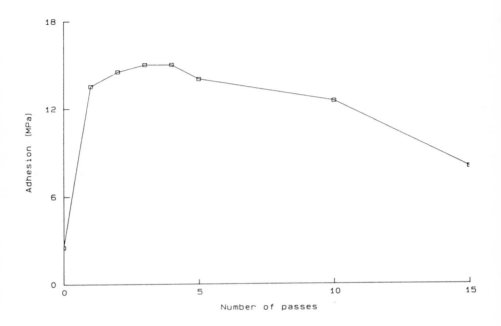

<u>Figure 3</u> Effect of number of passes of blast stream on coating
adhesion

and the plane of the substrate surface) and plasma spraying at the same angle of 90 degrees. Experiments were also carried out to investigate the influence of this angle on adhesion. Samples were prepared by both grit blasting and plasma spraying at 45 degrees as well as at 90 degrees. The results in Figure 4 show that grit blasting at 45 degrees followed by plasma spraying at 90 degrees led to a deterioration in adhesion. However, plasma spraying along the same direction as the 45 degree grit blasted substrate increased the adhesion, possibly due to improved penetration of the alumina coating particles into the valleys of the substrate profile and enhanced mechanical keying under a normal (90 degrees) test stress.

(b) Surface topography

The results show that the grit blasting conditions have a pronounced effect on the coating adhesion and that this can be broadly understood in terms of roughening of the substrate. Nevertheless, a more precise description of the substrate profile is likely to provide a clearer understanding of the critical mechanisms.

The average roughness, R_a, is equal to the area of the peaks (solid) above the mean line plus the area of the valleys (void) below the line divided by the length, as illustrated in Figure 5. R_a provides a measure of the average height or amplitude of the peaks in the profile. Figure 6 shows a clear correlation between R_a value and coating adhesion, which is consistent with qualitative practical experience. The R_a value is an average amplitude parameter, however, and provides no information on the slope of the peaks nor their spacing and does not distinguish between a peak or a valley.

The average slope of the peaks, Δa, gives a measure of their sharpness as illustrated in Figure 7. Figure 8 suggests that the coating adhesion may increase with increasing Δa value. In order to gain more insight into this effect, a grit blasted substrate was lightly polished, which markedly affects the peak slope as shown in Figure 9. Figure 10 indicates that polishing adversely affects coating adhesion, which supports the view that average peak slope has a significant effect on adhesion. In physical terms, the effect of peak slope could be related to the tendency for incoming molten droplets from the plasma jet to be punctured and pinned more readily by sharp peaks.

The average spacing between peaks, S_m, is the number of peaks over a given height divided into the evaluation length (Figure 11). The results obtained to date are presented in Figure 12 and suggest that there may be an optimum peak spacing of approximately 200μm. An optimum peak spacing could possibly be related to the dimensions of a molten droplet as it spreads out to form a splat on impact with the substrate. However, there is significant scatter in the data and more work is required for confirmation and to establish its physical basis.

A difficulty in the analysis of the surface profile data in this investigation, in which all the surfaces were treated with the same manufacturing process, is that many of the parameters are

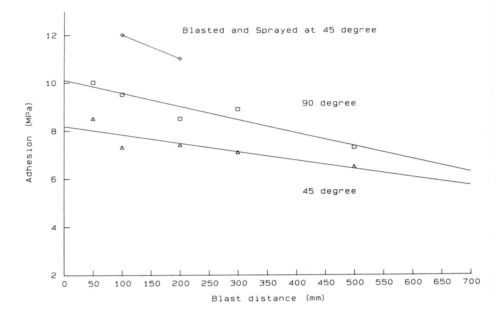

Figure 4 Effect of blast angle on coating adhesion

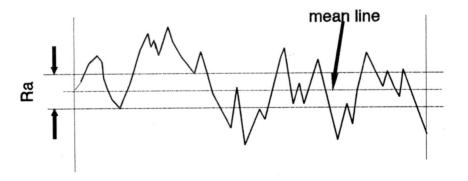

Figure 5 Definition of R$_a$

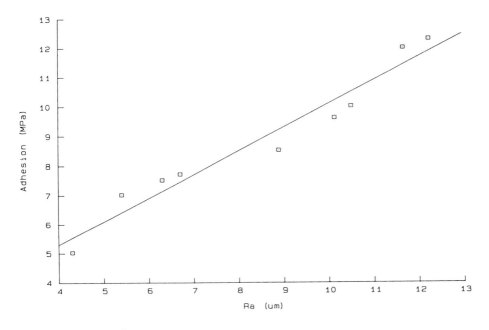

Figure 6 The effect of R_a value on coating adhesion

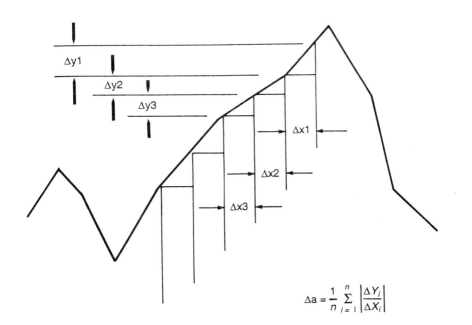

$$\Delta a = \frac{1}{n} \sum_{i=1}^{n} \left| \frac{\Delta Y_i}{\Delta X_i} \right|$$

Figure 7 Definition of the average peak slope, Δa

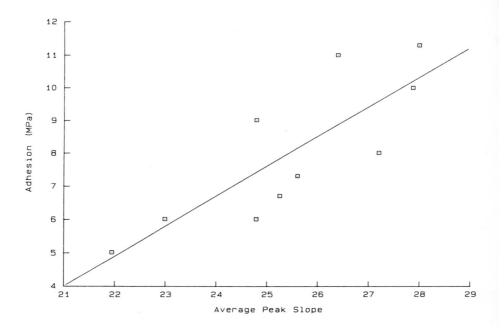

<u>Figure 8</u> Effect of average peak slope on coating adhesion

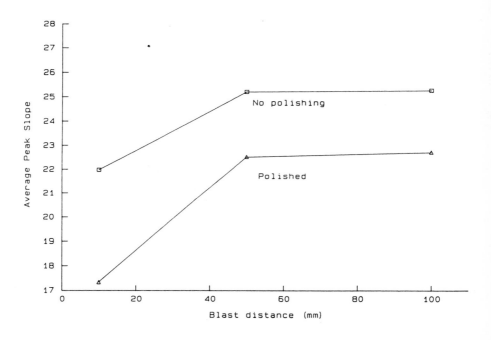

<u>Figure 9</u> The effect of light polishing of the grit blasted
substrate on the average peak slope

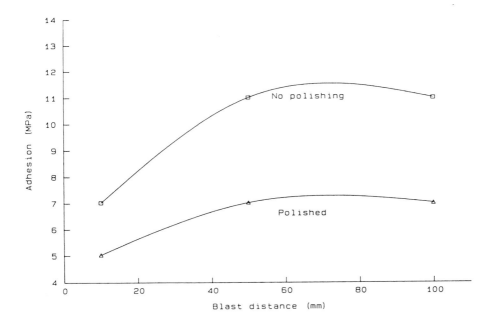

Figure 10 The effect of light polishing of the grit blasted
substrate on coating adhesion

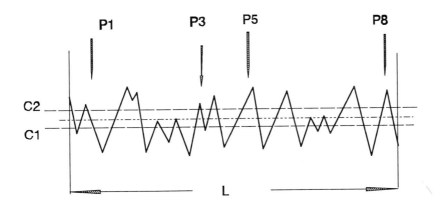

Figure 11 Definition of average peak spacing, S_m.
S_m is the evaluation length divided by the number
of significant peaks, where a significant peak has
a peak above C2 and adjacent valley below C1

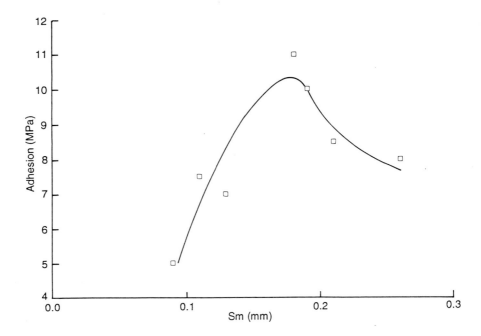

The relationship of average peak spacing, S_m, and coating adhesion

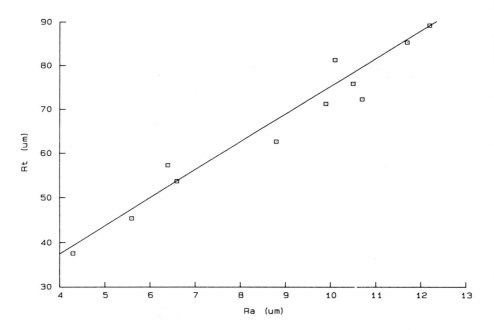

<u>Figure 13</u> The relationship between R_a and R_t values

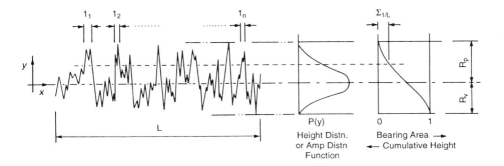

Figure 14 Definitions of amplitude distribution function and
bearing area curve

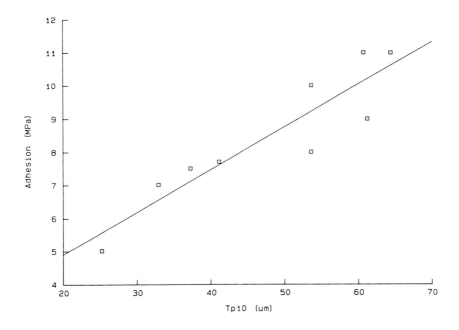

Figure 15 The relationship between T_{p10} and coating adhesion

The Trace Of A Surface Manufactured Using A Symmetrical Tool

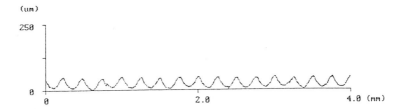

The Bearing Area Curve For The Above Surface

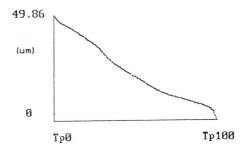

The Characteristic Asperity For The Above Surface

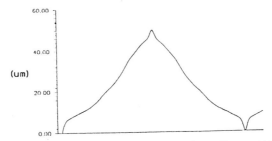

Figure 16 Profile trace from a turned surface with the
         associated characteristic asperity (after
         reference 2)

interrelated. For example, Figure 13 shows that the R_a value closely correlates with the R_t value (the maximum peak to valley height over the evaluation length). Under these circumstances, it is useful to relate the correlations to physical mechanisms and to employ wider geometrical representations of the profile.

A useful geometrical representation of an engineering surface is the bearing area curve and its derivatives. It can be generated by slicing the profile parallel to the mean line from the highest peak downwards and plotting the fraction of solid intercepted over the total slice length (or area in three dimensions) as illustrated in Figure 14. The fraction of solid varies from 0% at the highest peak to 100% at the deepest valley. The shape of the bearing area curve is a valuable descriptor. A fairly horizontal bearing area curve indicates a flat surface, which would provide even load distribution and make a good bearing. A predominantly sloping bearing area curve indicates a peaky surface with many outstanding peaks. The fraction of solid at any particular height is called the bearing area of material ratio, and more specifically the t_{p10} value refers to the depth from the highest peak at which 10% of the slice length is solid. The t_{p0} represents the topmost peak and t_{p100}, the lowest valley. A large t_{p10} value indicates a peaky profile and a low value, a fairly flat surface.

The relationship between the t_{p10} value and coating adhesion is given in Figure 15. The results suggest that peaky surfaces generally favour adhesion, possibly because they tend to pin and anchor most effectively the incoming droplets of coating material from the plasma jet.

(c) Asperity shape

The topography has been related to adhesion in this paper up to this stage by using a series of numerical parameters one at a time. This has limitations and an alternative approach is to attempt to encapsulate more of the surface characteristics in a visual representation of the average asperity shape.

A highly deterministic process like turning produces a regular surface profile that can be represented geometrically so that an average asperity can be readily drawn and visualized. However, grit blasting is a random cumulative unit event process which produces a randomly shaped surface profile. With such a wide variety of peak shapes and sizes, it is difficult to specify the surface. This difficulty in visualizing the surface in a simplified manner inhibits an understanding of the way the surface performs in practice. A method of treating this problem is to consider a surface constructed of asperities based on the characteristic form of the bearing area curve (2).

When the bearing area curve is plotted with the vertical axis equivalent to the profile height, not only is the form specified but also the magnitude of the height. In this form, the bearing area curve represents half a theoretical asperity. When the bearing area curves are placed back to back, they define a characteristic asperity. This is best illustrated by Figure 16 (after Furze and

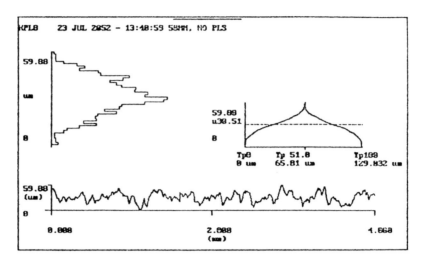

(a)   Non-polished, Adhesion 11.2MPa

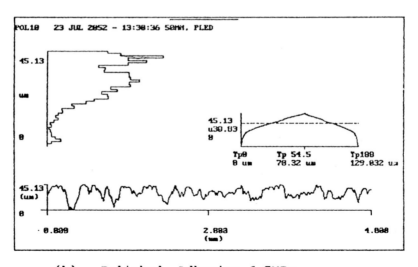

(b)   Polished, Adhesion 6.7MPa

Figure 17 Characteristic asperity forms for the lightly polished
and non-polished grit blasted surfaces

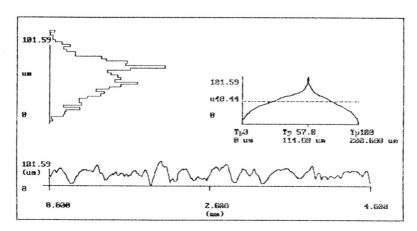

(a)   Distance 50mm, Adhesion 11.4MPa

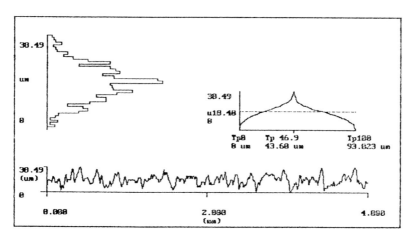

(b)   Distance: 500mm, Adhesion: 4.8MPa

Figure 18 Characteristic asperity forms as a function of blast
distance

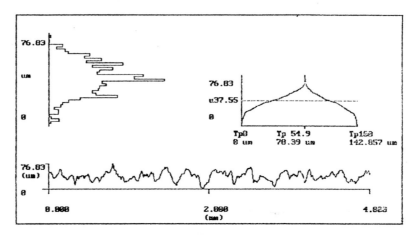

(a)   P-type, Adhesion: 10.3MPa

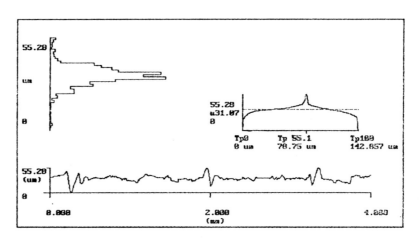

(b)   S-type, Adhesion: 2.25MPa

Figure 19 Characteristic asperity forms of a suction-operated
machine and pressure-operated machines

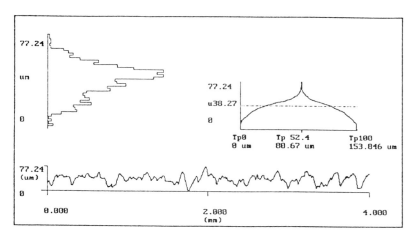

(a)   Angle: 90°,   Adhesion: 10.82MPa

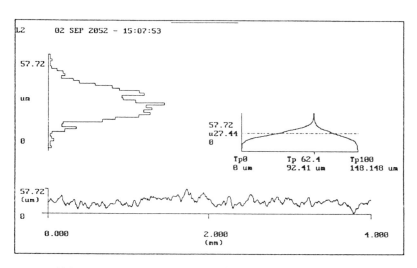

(b)   Angle: 45°, Adhesion: 5.4MPa

<u>Figure 20</u> Characteristic asperity forms for substrate grit blasted
at (a) 90 degrees, and (b) 45 degrees

Griffiths (2)) in which a schematic profile and the corresponding bearing area curve of a turned surface are shown. The bearing area represents half an asperity if the horizontal scale corresponds to half the asperity spacing, in this case, half the feed-rate. Any surface can thus be represented by a series of back to back bearing area curves producing a characteristic asperity. Included on this diagram is the tool shape as obtained by metrology measurements on the same scale. Deviations between the theoretical asperity form and the characteristic form are due to the minor variations and disturbances that occur during turning, such as deposits, tears, laps and folds, which are the non-deterministic aspects of a turning operation.

This approach is particularly useful in the treatment of grit blasted surfaces, since it enables a visual representation of a highly random and disordered profile. The resultant average or characteristic asperity form is clearly a significant simplification of reality, especially in view of the wide variations about the mean, but is nonetheless a valuable aid in understanding the role of the surface profile in practice.

Figure 17 shows the characteristic forms for the lightly polished and non-polished grit blasted substrates, and illustrates the flattening of the asperity peaks caused by polishing. Figure 18 reveals that decreasing the blast distance produces deeper valleys with wider spacing. Figure 19 indicates how the more powerful pressure-operated grit blasting machine generates much deeper valleys than that with the suction-operated machine.

The characteristic asperity presented above is derived from a single bearing area curve and therefore must be symmetrical. Many asperities are, of course, asymmetrical in practice and this can be accommodated by the following procedure (2). If the profile height data are put into two arrays representing respectively the left-hand peak data (the 'up' side of the peak going from left to right) and the right-hand peak data (the 'down' side of the peak), then two bearing area curves can be obtained which together more accurately describe the characteristic asperity. To obtain the two bearing area curves the array of profile heights is interrogated sequentially and if the height of point (n+1) is higher than that of point n, then the gradient is increasing and the data point corresponds to the left-hand asperity information, and vice versa for the right-hand. Each array will produce its own bearing area curve and when they are placed back to back will provide the possibility of an asymmetric asperity. The asperity length again corresponds to the average peak spacing.

Figure 20 shows data for substrates grit blasted at 90 degrees and 45 degrees to their surfaces. The surface blasted at 45 degrees reveals the expected asymmetry indicating the potential of this method in the analysis of random surface topographies.

    4.   CONCLUSIONS

1.   There is an optimum blast distance for maximum coating adhesion.

2. The coating adhesion increases with increasing blast pressure over the range used.

3. There is an optimum number of passes to achieve maximum coating adhesion and overblasting results in a substantial deterioration.

4. For plasma spraying at 90 degrees, grit blasting at 45 degrees produces lower coating adhesion than that at 90 degrees. However, improved adhesion results if the substrate is grit blasted at 45 degrees and then plasma sprayed in the same direction.

5. The average roughness (R_a), average peak slope (Δa) and the t_{p10} value all act to increase the coating adhesion under the conditions studied.

6. The results suggest an optimum peak spacing for maximum coating adhesion.

7. The use of an analysis to generate a characteristic asperity form for a given profile is shown to have potential for the study of grit blasted surfaces.

ACKNOWLEDGEMENTS

The authors would like to thank the ACME Directorate of SERC, Guyson International Ltd. and AEA Industrial Technology, Harwell Laboratory for permission to publish this paper.

REFERENCES

1. M.G. Nicholas and K.T. Scott, Surfacing Journal, 12 (1981) 5.

2. D.C. Furze and B.J. Griffiths, To be published in Topography.

Conference Photographs

The following plates show speakers, guests and delegates at the 3rd International Conference on Advances in Coatings and Surface Engineering for Corrosion and Wear Resistance, and the 1st European Workshop on Surface Engineering Technologies and Applications for SMEs. Both events were organized by the Surface Engineering Research Group of the University of Northumbria and took place between 11-15th May 1992 at Newcastle upon Tyne, UK.

Contributor Index

This is a combined index for all three volumes. The volume number is given in roman numerals, followed by the page number.

Subject Index

This is a combined index for all three volumes. The volume number is given in roman numerals, followed by the page number.